D1265401

MEMOIRS OF AN ADDICTED BRAIN

Memoirs of an

ADDICTED
BRAIN

A Neuroscientist Examines His Former Life on Drugs

MARC LEWIS, PhD

PublicAffairs
New York

PublicAffairs books are available at special discounts for bulk purchases in the
United States by corporations, institutions, and other organizations. For more
information, please contact the Special Markets Department at the Perseus Books
Group, 2300 Chestnut Street, Suite 200, Philadelphia, PA 19103, or call (800)
819–4145, ext. 5000, or e-mail special.markets@perseusbooks.com.

Cataloguing-in-Publication data for this book are available from the
Library of Congress.
ISBN (hardcover): 978-1-61039-147-4
ISBN (e-book): 978-1-61039-148-1
LCCN: 2011944105

First Edition

10 9 8 7 6 5 4 3 2 1

For Isabel,
who never lost confidence in this book or its author

CONTENTS

Introduction   1

PART ONE: THE TABOR CHRONICLES

Chapter 1. Changing state   7

       2. Giving up control   27

       3. Into the fire   43

       4. Dopamine and desire: A romantic interlude   59

PART TWO: LIFE AND DEATH IN CALIFORNIA

Chapter 5. Pulling out the stops   71

       6. Psychedelics, sex, and violence   93

       7. A psychedelic finale: Cops and angels   115

       8. Heroin, the Heap, and the sleep of the dead   127

       9. Getting down   146

PART THREE: GOING PLACES

Chapter 10. Travel broadens the mind   169

        11. Consciousness lost and found   193

        12. The opium fields   213

PART FOUR: IN SICKNESS AND IN HEALTH

Chapter 13. Night life in Rat Park   239

        14. Crime and punishment   255

        15. Healing   290

Epilogue   302

Acknowledgments   307

Endnotes   309

INTRODUCTION

WE ARE PRONE TO A CYCLE of craving what we don't have, finding it, using it up or losing it, then craving it all the more. This cycle is at the root of all addictions—addictions to drugs, sex, love, cigarettes, soap operas, wealth, and wisdom itself. But why should this be so? Why are we desperate for what we don't have, or can't have, often at great cost to what we do have, thereby risking our peace and contentment, our safety, and even our lives? Why are we so moved by our addictions, either succumbing to them or spending our energy fighting them? This book shows how the fundamental workings of the brain are at the root of the problem, but from a unique vantage point. I use the events of my own life as a springboard to the addicted brain. In fact I'm an expert on the matter, from inside out and outside in, because I'm a drug addict turned neuroscientist.

In some ways, drugs are just another addiction. More powerful than many, more harmful than most, but just another pot of gold at the end of the rainbow. Yet drugs are also unlike any other addiction, more revealing than any other addiction, because they ignite the neural flush of well-being directly, without the requirement of any particular experience or event. Drugs trick the brain into dispensing the neurochemicals of reward, or they mimic those chemicals themselves. Drugs provide a shortcut. They talk to the brain in its own language—the language of dopamine and peptides, neuromodulators and receptors. So drugs can teach us a lot about the brain, and what we know about the brain can teach us a lot about addiction. Neuroscientific research on addiction has progressed enormously in the last twenty years. But it doesn't go all the way. It lays out the pieces of the puzzle but doesn't connect them, because it ignores the actual experiences that turn real people, motivated by hope as much as hedonism, resolve as much as indulgence, back into addicts, again and again.

This book brings the brain and human experience together, by telling the story of my descent into drug addiction, interspersed with lessons on the brain and its workings drawn from contemporary neuroscience. I follow the thread of adventures that began in a New England boarding school, when I was fifteen years old and despondent, a thousand kilometres from my home in Toronto. From there I moved to Berkeley, California, in the heyday of the drug movement, and upgraded my flirtations with cough medicine to an infatuation with LSD and then heroin. Interspersed with university life in Berkeley, I spent two years in Asia, where I joined medics sniffing nitrous oxide in the Malay jungle, bought heroin direct from the factory in Laos, and became a regular in the opium dens of Calcutta. I moved back to Toronto, married a woman I couldn't talk to, began stealing drugs from the psychology lab, then from medical centres, got clean, got divorced, fell off the wagon again, and ended up working in a mental hospital where the howls of the crazy people drove me back to crime. I finally got arrested and convicted, and then came the tortuous road to a lasting recovery. At the age of thirty, I traded in my pharmaceutical supplies for the life of a graduate student, eventually becoming a professor of developmental psychology, and then of neuroscience—my field for the last twelve years.

Now I study the brains of children who get into trouble, using the electrical signals from their scalps to explore what's going on underneath. But I often recall that I used to be one of those kids—that, no matter how many scientific conferences I attend, I will always be one of those kids.

Each chapter introduces a new drug experience, or a new stage of drug addiction, as the central theme of an episode from my own life and, sometimes, the lives of lovers, seekers, and sinners I met along the way. And each of these experiences is shown to emerge from a particular brain system, neurochemical flow, or synaptic process. Through the interplay of lived experience and neural activity, the addicted brain is revealed, and addiction is shown to be a basic vulnerability of the nervous system itself.

The episodes recounted here are all factual, to the best of my ability to remember them, though some are blended to maintain the pace, and names and locations are sometimes disguised. I could not invent a story as strange and frightening as the life I lived as an addict. The dialogue is a best approximation of actual conversations, with some snippets burned in my memory and others deduced from more general recollections. Over twenty volumes of journal entries helped me recall what life was like in my teens and twenties. Finally, the neuroscience in these pages, though simplified and accessible to the nonexpert, is up to date and accurate. Many "brain books" pass on findings from twenty years ago, or blur the details to make them more digestible. Here the details are precise and the data as current as last year's journal articles. The brain is incredibly complex, but it is not beyond understanding.

• PART ONE •

THE TABOR CHRONICLES

1

CHANGING STATE

The first time I got drunk was with Damien Tennant. It was a March night and freezing cold. I had been at Tabor less than six months, and depression was now a palpable daily companion, a disease to be managed. There were good days and bad days, but the good days were just tolerable and the bad days nearly did me in. I hated this school. It was bigger than me, it was stronger than me, and it seemed as natural a part of the New England landscape as the rocky coves and stands of maple. I didn't belong here. I had got here by mistake. My first two years of high school had been at a nice, normal suburban Toronto school a few blocks from my home. I might not have been the most popular kid in class, but nobody seemed to actively dislike me. I had a few friends. I got invited to a few parties. I had a girlfriend for a couple of months. And I could go home at night. That was the main thing. Here at Tabor, there was no home to return to at the end of the day—except the dorm. And I hated the dorm. I hated every whitewashed board, every hissing radiator, every creak in the polished hardwood floors. I hated my room, I hated my roommate. I hated the guy next door. And I hated my proctors, the senior boys whose job it was to supervise us and care for us while helping induct us into this bizarre paramilitary culture.

Para-naval, actually. Tabor was a para-naval academy, but that didn't mean much to me at the time. The naval terminology seemed a bad joke, an effort to reenact the Hollywood heroism of World War II. We would sometimes see men who looked like admirals strolling the pathways with the headmaster. Gold stripes on dark blue cuffs. There were special prayers at vespers, navy uniforms to be ironed and worn a few weeks a year, drill sessions with guns and a marching band. And yes, we were located on a body of water and there were a lot of boats: sailboats, rowboats, crew boats, a schooner out in the harbour. It was the boats that had attracted me in the first place, the previous spring in Toronto.

"Marc, we'd like to talk to you about your options for next year," my mother had said in her frank, slightly invasive tone.

"What options?" I was beckoned to the kitchen table, where I sat down with both parents. The table was covered with pamphlets from different private schools, most of them in New England.

"You seem a little bored with school," Mom continued. "And maybe a bit discontent overall?" Those penetrating eyes—hazel and clear with mildly arched brows—scanned me for an accuracy reading. They gazed at me steadily from her pretty, mid-thirtyish face—a face surrounded by sprayed-in-place hair that became increasingly blonde as the sixties wore on. My mother seemed to wield her own sixth sense. She could look into me and find things. As usual, I soon felt I must be hiding whatever it was she was looking for.

"I'm fine, Mom. Everything's okay." My dad sat beside her, hunched over and uncomfortable. His pleasant features, thin black hair, and solemn brown eyes would have been more at home at a meeting of the family-run leather business. Dad didn't usually get involved in family discussions with an emotional theme, if that's what this was. Maybe he was in the dark as much as I was. His

8

half-smile tried to make light of things. But this was sounding ominous.

"I know everything's fine," she said. "But remember we talked about boarding schools? And you said you might be interested? Remember, we'll be moving to San Francisco in two years' time. Going to an American school would save you a year of high school. You'd be able to go straight to college the year we arrive. So . . ." she smiled encouragingly, "I've done my homework and gotten these pamphlets from some very fine schools. And we thought," with a glance at my dad, "that you might want to look at them."

It all seemed exciting, though too far off to think about. So I looked at the pamphlets piled up in front of me. We looked together, and my parents pointed out this and that feature of this and that school. I tried to pay attention, but my thoughts wandered like stray cats. I was preoccupied—some sort of creeping disorientation at the very prospect of leaving home. And beneath that, intangible but potent, a sense of dread. Were they trying to get rid of me? Had I done something wrong?

Now Tabor was my world: there was nothing left to decide. Every morning I joined four hundred other teenage boys tramping along a wooden walkway between brick buildings in a cold mist, breakfast settling, the assembly hall looming, thinking about ways to avoid attention. Our headmaster, Mr. Witherstein, looked as though he'd stepped out of an old movie. He parted his hair almost dead centre, and he was grey and crusty like a venerable admiral himself. He waited for us at the lectern, beaming with unnatural enthusiasm. He delighted in the series of routine announcements that he would soon recite with such gravity. I took my assigned seat. Each seat back held a short wire rack containing a hymnal. I reached toward mine automatically, on cue. We opened our hymnals together, the rustle of pages filling the hall, and we began to sing.

A mighty fortress is our God! A bulwark never failing. God had never appeared to me as a fort or a bulwark. Maybe that was because I was Jewish. Yet I didn't mind the singing. It numbed me. There was a comforting anonymity in being a part of this mass, this sea of boys packed row after row in a heated auditorium. I was safe for now.

Back at my dorm, it was survival of the fittest, and I wasn't all that fit. A pecking order had consolidated in the early months, and I was pretty close to the bottom. At first my roommate, Todd, was the victim of some pretty vicious teasing by just about everyone, but especially Bob Moore, the handsome giant who lived in the room next door. Moore was a front-line football player, though still a junior, and for that alone he was universally admired. Todd put some sort of lotion on his face at night, giving it an awful vampiric white-ness, all the more disturbing with his black stubble poking through. He was soon known to one and all as Madame Butterfly. I was sym-pathetic at first, but I didn't like him much. He was caustic, whiny, and sour. I tried to like him. I tried to be nice to him. But I secretly hoped that his victimization would keep me safe, give me some breathing space. Meanwhile, Moore had risen rapidly to power. His own roommate, Randy, was an absurd-looking, gangly outcast, with protruding ears, perfectly designed by nature for the role of village idiot. And he was Moore's roommate! I felt sorry for Randy. What tor-ments must he suffer? But I was relieved that Moore was surrounded by victim material. Although I knew that was selfish, I needed it to be so. I needed the playing field tilted in my favour.

And then, to my horror, it tilted the other way. The dominance hierarchy heaved a final time and Randy and Todd ended up Moore's lieutenants, his slaves. Randy brought him anything he asked for. He literally stood around, waiting for orders. Todd found his niche fawning over Moore from the sidelines, grinning at his savage jokes and heaping ridicule on his victims.

I was next in line.

"Hey Lewis!" Moore smiled at me almost warmly from the doorway connecting our rooms. His broad, manly features beamed with good humour.

"Yes?"

"Come on in. Join the party."

"Okay." I couldn't refuse.

"Did you see what happened to my dresser?"

"No . . ."

"Take a look."

"I don't see anything."

"Come closer." He seemed so inviting. I wanted to believe that I was really being included. But I didn't see anything unusual.

"Closer!"

"I don't see . . ."

"Bring your head right up to the edge here, ya dummy. You have to look at it from just the right angle." I did as I was told, moving my face to within inches of the dresser surface. There seemed to be a pool of water there.

"I see some water. How did that happen?"

Then smash! His palm came down hard on the little puddle; my face was soaked. And whatever was running down my cheeks stung my eyes, which started instantly to tear. Aftershave? Hydrochloric acid? Liquid shame? My tears were the worst part.

"How did that happen?" Moore echoed in mincing tones. "How did that happen?" His voice rose with incredulity while he danced around the room. "Did it make you cry, Lewis? Poor baby, we'd better call your Mama," and they all erupted into laughter, Moore, Randy, Todd, and another guy. I looked through tears at Moore's grinning hyena mask, curling now with contempt, and I wanted to slip through the floor. *What was wrong with me?!*

11

I began to avoid my dorm whenever I could, to synchronize my comings and goings with those of the other boys, so Moore couldn't get me alone. I made a couple of friends and that helped. There were other misfits who had no interest — and generally no place — in the jock-dominated hierarchy. By Christmas I spent a lot of time with Schwartz and Burton. Joe Schwartz was a junior like me, but he was leather-tough in some unique way. Self-sufficient and really smart. Burton was just a big teddy bear, gruff and mischievous, whom nobody could dislike. I got to know their friends, Gelsthorpe, Perry, and Norris, all seniors. These were the guys who would graduate at the end of the year, and they all lived in another dorm, Pond House. I spent a lot of time at Pond House, and I wished I could switch. But at night I had to return to my own dorm, with its sixteen boys, often creeping up the back stairs to avoid notice. Sometimes, before going to my room and facing Todd, I'd drop in on Lawrence Carr, one of the two black kids at Tabor. That's two out of four hundred. The other one, Lavalle, was despised by all: he wore his bitterness like an armband. But Carr stayed above the fray and was left alone. In my eyes he was magnificent.

"How do you handle it?" I wanted to know.

"I take it all for granted," he replied. His smile was gentle, chiding. "What did you expect, man? Tabor is no temple of brotherhood. It is exactly what it appears to be."

"Yes, but . . ."

"Just wait it out. Keep your own council."

It seemed easy for him.

When I finally learned how to avoid Moore's displays of goodwill, trouble came from a guy named Wiley who lived across the hall. Passing his door became an agony of nerves.

"A Jew! A Jew! Excuse me! I mean ah-choo! Bless me! Hey Lewis, did you hear someone sneeze?" His voice rang with feigned bafflement. Wiley was rotund and greasy, with small eyes and a blistering tongue. He sat chinless, unmoving, on a chair well inside his room, opposite the door, waiting for fresh bait to pass within range without his having to move from his lair. I was bait. Defenceless bait.

"Don't let them know they're getting you down," my mother advised on the phone. "If they don't get a reaction, they'll stop." That was probably the worst advice I'd ever got from anyone. Or maybe I just didn't know how to implement it. I couldn't imagine how I might pretend that it wasn't getting me down. It crushed me! Wiley could see my face as I hurried by or see me looking down, avoiding his gaze.

But I soon got more potent advice from an unlikely source, a boy named Miles. He was a loner like Carr, and confident like Carr—in fact contemptuous of the scurrying cruelty around him. I sat at the foot of his bed one night, looking at unfathomable pictures of his enormous home in the Deep South.

"Lewis," he asked, leaning regally against the headboard, "why do you let Wiley pick on you?" He studied me carefully, trying to discern the peculiar defect that must be there.

"I don't . . . mean to. What choice do I have?"

"What choice?" He snorted with mild disgust. "Lewis, your problem is that you have no common sense. You're an intelligent kid, but you don't know how to use it."

"What should I do?" If there was some workable trick, I wanted badly to know it.

"Look at Wiley. Does he have any weaknesses?"

"Well, no. He's got lots of friends in fact."

"Yeah, but is there anything about him that's . . . that he's not proud of? That he doesn't want his friends to see?"

"Just that he's fat, I guess."

"That's right. He doesn't like being fat, does he?" as if to a six-year-old. "So you could get him just as bad as he gets you."

Miles proceeded to train me in the art of verbal swordplay. I practised daily, in pretend showdowns, saying "Hey Wiley, you're so fat, your mother needed explosives to get you out. Hey Wiley, you're so fat, you can't reach your asshole to wipe your shit. Which is why you stink so bad. Hey Wiley, the reason you never come out of your room is because you can't fit through the door." I couldn't imagine actually saying such things aloud. Until one weekend morning, incredibly, I dived off the deep end and blurted one of these haikus right at him, right in through his door as I passed by. His contemptuous smirk collapsed into outrage and — could it be? — a touch of fear. As if he'd been stung by an ant.

Wiley remained manageable as long as I flung the occasional caustic remark back at him. He was reduced to snickers and glaring. But I couldn't fight the proctors. They had ultimate authority, ultimate confidence, and they seemed to relish punishing me. At one point I had to shine their shoes and leave them neatly by their doorpost. Every day for two weeks. And much to the delight of Moore and Wiley. That was my punishment for coming late to breakfast one morning. Another week I had to make their beds — perfectly — for the crime of having left mine unmade in my rush to be on time. That's how I learned about promptness and order. But worse than the punishments was the scorn that went with them.

"Not good enough, Lewis. Do it again."

"What are you doing here anyway, Lewis? Did they kick you out of Canada for being a fag?"

"Hey, these shoes shine! You see, you *can* do something right."

Why did they torment me? There must be something wrong with me, but what? I looked intently at myself in the mirror. All I saw was a regular-looking fifteen-year-old kid with wavy brown hair, some dark stubble spreading about unevenly, my mother's hazel eyes, an okay nose—a bit Jewish, but not unseemly. My mom even said I was handsome, but what else would she say? And anyway, I wasn't the only minority. I wasn't the only Jew or the only foreigner, though there weren't many of us. A dark rumination spread week by week. It bubbled to the surface on the way to and from classes, during mealtime, while lying on my bed doing nothing, a sporadic flow of sulphurous toxins. Snatches of dialogue laced with anxiety, shame, and dread. Spirals of self-criticism. Why can't you get it right? Why can't you *learn*? What the hell's wrong with you? Why are you such a wimp? A running tirade of self-blame like an invasive radio station. And why not? I was far away from home, I was not very well liked by my immediate associates, and my parents seemed to have lost me on their radar.

One night just before Christmas break, I was so desperate that I knocked on the door of the master who lived below us, Mr. Wharton, whose house this was. He and his wife and two attractive daughters inhabited a large apartment on the ground floor, a real home with a living room and a kitchen and bedrooms. But the masters' homes were out of bounds. In fact the masters were pretty much out of bounds. We faced them in the classroom, the sports field, the dining hall. We sang with them at vespers. And that was it. Now it was ten at night. A massive Christmas tree dominated the room before me, but Wharton blocked it like a football lineman. His eyes were hard, cold, and spookily bright. He leaned up against the threshold, crossed his arms, and asked me, "What's the problem?" A tale of woe came spilling out of me quickly at first, but it soon began to sound ridiculous, pathetic, cowardly, even to my ears, and trickled to a stop. At which point Wharton smiled in some pseudo-fatherly way, his breath spiked with vodka, and reassured me that I would be just fine. The boys could be a little frisky sometimes, but they were just being boys. I was still new. And I was raised . . . differently. I'd toughen up soon.

Instead of toughening up, I spent more and more time at Pond House. Only five boys lived here, but others came to visit regularly. Pond House was a haven for Tabor's outcasts—failures at almost everything except writing, as demonstrated through longish existential essays that we hoped would keep our English master from sinking into despair. On a typical Saturday afternoon in February, I climbed up the side stairs to the growing whine of a Dylan record and bits of conversation trickling from Perry's room. I pushed the door open and a warm wave swept over me. There were friendly faces all around. Guys were sitting on desks, shelves, dressers, and beds, chatting about the stupidity of Tabor culture and their hopes for rebirth next year. I sat on the floor. Huge paper flowers poked out of holes in the wall. An iconic Allen Ginsberg smiled down from a poster above a bed. Coke cans, mostly empty, were scattered about the room, and peanut butter sandwiches passed from hand to hand. I felt safe. Later, when no one could agree on what record to play, we grabbed our coats and ran down to the shore, right at the foot of the snow-patched lawn. We climbed along the rocks lining the harbour, then took turns running out across the ice, way, way out to where you could see the dark water just underneath you. You could feel it pressing upward against this fragile membrane, waiting to engulf you if you chose the wrong path. We cheered each other on, crazily, with no concern for life or death, lost in the ecstasy of escape.

I was outside myself at last. If only I could stay there.

Teenagers spend an enormous amount of effort creating themselves. They find themselves in a world where social comparison means everything, far more than the popularity contest that starts at age five. It's now a sudden-death tournament. When you lose, it's your whole identity that takes the hit, with no parents handy to put you back together. There are so many ways to be uncool, to find yourself on the outside, to lose hard-won social ground. The buttresses of character constantly need repair and rebuilding by dint of hard work and excruciating self-examination. It's no accident that adolescence is the period of greatest psychological vulnerability. In the years from thirteen to sixteen, every emotional disorder climbs in incidence from a flat line to a rising slope. Depression, suicide attempts, anorexia, conduct disorders, and of course substance abuse. Adolescence is a fragile time, when the rapid rate of self-renewal leaves patches of wet mortar, joints that endanger the whole structure. Where the fault lines touch the surface, exposing the core to infection and collapse.

By the time I got to Tabor, my newly minted persona was just starting to harden. It had elements of manly fortitude, humour—I could make people chuckle—and what I hoped was an understated but keen intelligence. I was a smart kid, likeable, generous, easygoing, stalwart, and brave, like my TV heroes, but with a winning sensitivity that girls would appreciate. At least that's who I wanted to be, and I was working on it. Like other kids my age, the big job was to keep juggling the pieces until I could catch them one by one and stick them onto an accumulating selfhood.

It's a tough job, but I was managing. Until I got to Tabor. And that's where I lost a critical part of the foundation, a concrete floor, needed to hold up the rest. What I lost was a sense of security. There was no safety in my world, no home, no peace. Currents of anxiety fractured every mood. My bed was my only haven, and even that shelter was violated at regular intervals by Todd, fresh and white-faced, a grinning whore evicted for the night from Moore's den of cruelty, come to snip at me before I could pretend to be asleep. It didn't take long to realize that Tabor was a colossal mistake. A wrong turn. And somehow I thought it was my fault, which meant I had to bear it. I had to go through with it. I tried not to think about why I was sent here, but rather to deal with the challenges in a manly way. Day by day. To prove to everyone that I was strong enough.

But I wasn't strong enough. Depression gutted me, and I had no idea what to do about it. I had no manual, no prescription for dulling the ache of homesickness compounded by the unpredictable snares around me. Instead, perhaps inevitably, despair finally pried some rebellious urge out of me. By the last grey months of winter, I was ready to protest.

Most boys who lose themselves on the path to adulthood start to engage in "antisocial behaviour," and that's what happened to me. For the first time in my life, at least in any serious way, I was ready to break the rules. We were expressly forbidden from leaving the property at night. The forest was out of bounds, day or night, and as for alcohol . . . ! You might as well desecrate the Bible or the flag as get caught with booze. So when Damien Tennant offered to get me drunk, my nervous system crackled with excitement.

We were standing in line in the cafeteria, waiting to collect our trays of overcooked meatloaf, beans, and potatoes, when he whispered, "I've got booze. Meet me tonight in the woods and I'll share it with you." I didn't know him well, but he looked lonely and a bit forlorn — bored, distracted, perhaps friendless — and that warmed me toward him. His offer seemed to come out of nowhere, but I didn't care. "Sure! Just tell me where."

He was as good as his word, standing at the entrance to the forest just across from Lillard Hall, back from the road to avoid discovery. It was long after dark. I half-ran to meet him and we quickly ducked under the trees. We were surrounded by forest in moments, but we kept going, slogging through the wet snow on its unmade bed of frozen leaves. We followed a vague path, a swath of empty space defined by a slightly lighter shade than the black branches on either side. We found a small clearing with fallen trunks that served as benches and sat down, by mutual consent. Damien pulled a bottle of scotch from his pocket. We must be miles from other humans. Our crime would be invisible, inaudible, unknowable.

We have nothing to say to each other. So, without formalities, Tennant screws off the top of the bottle, takes a slug, tries not to grimace, and smiles his fake smile. He passes the bottle to me and I drink. It tastes awful, but the burn of its passage warms me and I like that. Back to Tennant, back to me, back to Tennant, back to me. And then I forget to pass it back. My gaze remains stuck on a clump of black matter at the edge of our clearing, and that fixation feeds back to my sense of what's going on: What are we doing here? We're getting drunk—that's the idea. And now my thoughts gather to take stock of the changes that creep in from every sense. The silence deep and ringing. The darkness unpacking itself into tiles of light and shadow. I want to share this shift in reality.

"Damien, I think I'm feeling something."

"I hope so," he answers with vague sarcasm. "It'd be a shame if you didn't." But the anxiety that would normally arise from his taunt melts as soon as it starts.

"It *would* be a shame, Damien. But luckily I do feel something. Quite a lot, actually."

He takes another gulp and grins, looking straight at me for the first time. "Luckily, I'm feeling something too. Luckily, we're both feeling something."

We burst into laughter. This is the funniest comment anyone could have made. We pass the bottle back and forth, grimacing, making dumb jokes, pretending we're anyone but unhappy high school students. And of course the alcohol comes on quickly. My body feels different: light and breezy, but then hugely clumsy when

I reach for the bottle too quickly. I'm startled by the change. My thinking is blunted, inarticulate, yet remarkably focused, like a zoom lens magnifying some meaningless piece of garbage. How is that possible?

I am cheerful and light-headed, and the cold begins to seem arbitrary, not threatening. In fact nothing seems threatening anymore. I am excited—giddy and gleeful with excitement. I am a special person doing a special thing: drinking. Getting sauced!

But I am also trying to figure out just what is happening to me. My thoughts bleed one into the next, joining up, converging. And yet I feel an accelerating clarity, a rising confidence. I feel special. Tennant and I share, as if by direct contagion, an unmistakable fervour, about being here, right here, right now, in this clearing, despite the cold, despite the dark blot of Tabor.

"Damien, I do believe I'm drunk!"

"Yes, Marc. I do believe you are. In fact, I do believe we both are."

"And I think I'd like to be even more drunk, if you wouldn't mind passing that scotch."

"For my old friend Marc, it would be a pleasure."

"It *is* a pleasure, Damien. I didn't realize what a great guy you are!"

We are talking utter nonsense and we are slurring our words, and that's just so funny that we break into gales of laughter. And now we are walking around in exaggerated marching steps, mimicking the moves we learned in drill, performing for each other in our little clearing. Navy men are we! The obvious idiocy of our lives now made tolerable, even joyous. But what amazes me most is that my familiar concerns seem to have vanished. I would normally obsess about the things I say to a new friend like Damien. Was that cool? Am I being too enthusiastic? Too friendly? Too anxious for approval? Am I annoying him? But all that stuff is just gone. There is no trace of hesitation. My words shoot out of my mouth, and

I watch them fan out across the playing field of our dialogue—carefree and confident. I'm just not worried about how the world, embodied now in the fuzzy form of Damien Tennant, will respond to me.

I've become bold.

We dance among the trees. And I feel undeniably, unambiguously happy. My excitement isn't just about being a bad boy out in the woods. I am excited because I am really happy for the first time in a long winter.

To what do I owe this magic, this alchemy? How does it work? Even now, as a groping teenager, I want to understand this metamorphosis. I have been deep in depression, the outcome of living in a circus of unkind acts. And suddenly I'm completely free of it. My mood has been knocked out of its basin and landed somewhere entirely different. I wake up, blink, and find I'm relaxed and agile, a cat on a fence looking back on my old haunts. How can that happen? What's the secret of changing how you *feel*?

The alcohol goes into my body, in through the lining of my stomach, into my blood, swishing around in its dark passages until it reaches the tollgate separating brain from body, the blood-brain barrier. And once past it, those alcohol molecules spread out from the brain's giant pipelines into smaller arteries, and then capillaries, and then to brain tissue itself, that incredible substance that is the source of all our experience. Brain cells, or neurons, suck those molecules in with their regular helpings of oxygen, and then . . . they change. We can follow these molecules down the rabbit hole, into the chemistry of the brain, into the neurons themselves. Because the key to getting high is a simple equation: brain change = mood change. The whole science of addiction starts here, where molecules from outside the body find communion with the cells we're made of. The brain that's become a creature of habit, nestling in its self-made cave of unhappiness, gets a wake-up call from a team of molecules born in a vat in Scotland. Somehow, and maybe not so surprisingly, we've learned to identify, to cultivate, to grow, distill, sell, and then consume gulpfuls of specialized but foreign molecules that have a particular affinity for these cells that matured inside our mothers' wombs. Thus begins the unholy marriage of intoxication. And in my case, it's a marriage of necessity.

Alcohol is the most boring of drugs, but it does what all drugs must do and it does it quickly and well: it changes the way you feel. And because you can get it off the shelf at the local store, it is the drug most abused by teenagers throughout the Western world. By the year 2000, most American high school seniors reported using alcohol within the past year, and 32 percent reported getting drunk at least once in the past month. This was the first time for me. The first realization that the internal landscape can shift like a tidal plane. In the blink of an eye. But certainly not the last.

The cerebral cortex is the most complex structure on earth—the crowning jewel of the mammalian nervous system—a convoluted carpet of cells that covers all the other parts of the brain and does the specialized work of thinking, planning, ruminating, imagining, taking stock of our surroundings, controlling our impulses, and then, at the right moment, propelling our muscles to act with skill and precision. The cortex contains much of the so-called grey matter: twenty billion neurons interconnected with one another in the most complicated circuitry. And this fibrous web of connections (between a thousand and ten thousand per neuron!) is where its work actually gets done. The cortex is a super-network of communicating neurons, babbling neurons, pulsing neurons, that constitute the physical substrate of who we are, what we think, what we do, what we experience. It is the resource that provides us with our skills, our abilities to analyze, judge, and make sense of the world. To daydream and ruminate. And it affords us a particular perspective, a mental framework, a "model" that coheres like a giant bubble on the surface of a roiling sea of emotions. This is home. Not the changing landscape of people and events around us. No, this electrical sea is where we really live.

The neurons of the cortex share information by releasing tiny amounts of chemicals to their neighbours, at the synapses, where the branching axon of one cell gives its messages to the dendrites—the receptor branches—of the cell next in line. Those chemical packets change the electrical charge of each recipient neuron, making it more likely to fire or less likely to fire in the next moment. In other words, they adjust its firing *rate*. That doesn't sound like much, but because neurons work in teams their contributions accumulate quickly, within milliseconds. That's how one neural neighbourhood influences the firing patterns of others, and those neighbourhoods, in turn, influence the firing patterns of others, and so on, in a tide of information that washes over the entire brain in less than half a second. That's the whole story. That's what happens every time you think something or feel something or move an arm to scratch something. How strange that the intricate symphony of our perceptions and actions can be reduced to changes in the firing rates of a bunch of trigger-happy cells. Can it really be that simple?

No. There are a few more tricks to neural communication, and the most important is the primal code, the machine language, that makes the whole thing work. The influence of one cell on the next takes one of two directions: it either makes it fire more often, called "excitation," or less often, called "inhibition." In fact, every cortical neuron can be identified as either *excitatory* or *inhibitory* based on this aspect of its job description. The molecules that actually do the work of crossing the synapses are called neurotransmitters, for obvious reasons. And drugs, including alcohol, very often transform the brain's firing patterns by altering the action of these neurotransmitters. There are two main neurotransmitters that correspond with the two modes of neural communication. Excitatory neurons send packets of *glutamate* across the synaptic channel—yes, glutamate, a derivative of the sugar molecule, glucose—while inhibitory neurons send packets of GABA—GABA is the short form for a name too long to pronounce. These two messengers have contrasting roles. Glutamate excites; GABA inhibits. That amounts to the yin and yang of the brain, the zeroes and ones of the fleshy computers we carry around in our skulls. But, just like the zeroes and ones of a mechanical computer, excitation and inhibition can be combined in configurations of enormous complexity. In a nutshell, excitation builds communities of neurons that activate each other, in self-amplifying waves of synchrony, like a spontaneous chant arising at a sports event. And these communities can form in less than a twentieth of a second, hence the speed of . . . well, of thought. Inhibition doesn't just reduce the flow: it also tunes neurons to become less responsive to their glutamate input, so that the waves of synchrony get differentiated, organized, tamed. Instead of one giant electrical tsunami—that would resemble an epileptic seizure, a mob scene rather than a chant—inhibitory neurons make sure that groups of neurons remain buffered from each other, limited in scope and focused on the task at hand.

What does alcohol do to this pulsing system of traffic signals? It enhances GABA transmission and squelches glutamate transmission. In other words, the inhibitory chemicals get boosted while the excitatory chemicals get hushed. A couplet of neural paradoxes. But of course, as a shivering sixteen-year-old, I don't have the manual at hand. That doesn't come till years later. I don't have a clue, as the scotch accumulates in my belly, that my cortex is starting to malfunction in two opposite ways at the same time. With each swallow, more of those little ethanol molecules find their way into my synapses, where glutamate and GABA are accustomed to crossing back and forth. *And everything changes as a result.* The sites of glutamate transmission become numbed and ineffective, so information flow is now sluggish, with big signals still getting through while small signals fade into static. That means I am noticing less, I am perceiving less, I am remembering less, I am feeling less. And that's a good thing: because what I've been noticing, remembering, and feeling lately have not been much fun. Meanwhile, GABA transmission gets an extra boost as it runs around shushing the electrical buzz, narrowing and selecting. It's GABA's job to fine-tune thought and perception, to clarify things, but now things are clear to the point of caricature. I reel with the inane certainty of the strident drunk. The stupid momentum that keeps me saying the same thing again and again: "Damien, this was a really great idea. A really great idea, Damien." Neuronal inhibition is not the same thing as social inhibition. It does not feel like suppressing something. Rather, it tidies up brain activity by dampening everything extraneous, by muting the fizzing uncertainty that's always there like background music. Now, with my GABA channels wide open, the background noise, the niggling uncertainty of normal cognition, gets turned off almost completely, leaving a glazed silence between flashes of thought. In other words, I am thinking about very little, but I am thinking about it with magnificent clarity. Thanks to the scotch bottle Damien holds out to me, I am thinking very clearly about almost nothing. And I like it! I look at Tennant with admiration and gratitude. He's the one who has given me this crisp, clear sheen of mental anaesthesia. And I like it! I am feeling better and better. I'm happy!

Yes, I'm happy. Because I like what I'm losing as much as what I'm gaining. The copiously textured network of ruminations about my flaws, my weakness, my penchant for being disliked, that unwanted radio signal of self-criticism, is finally silenced, as though I just drove into a deep tunnel. And it took this baptism of scotch to get here, to blanket the signal, so that I could finally, completely, thoroughly relax. All this, and the icing on the cake is a sense of belonging. I finally feel that I live here, in this body, on this planet.

Of course there was one additional payment to make. I woke up frozen, caked with vomit, on the forest floor, just as the sky began to lighten. And I was terribly sick for days after. So was Tennant. But despite this dreadful return to reality, I thought, someday I want to do this again.

Tabor days went on into April and May, and then June finally came, and with it the reprieve of summer vacation. I returned to Toronto not quite the person I was when I left. My parents found me uncommunicative, and indeed there was little I wanted to share with them. I wanted to be alone. I worked as a gardener in a cemetery, of all things. I read, I slept, I listened to my records. And I never knew whether my dad noticed that the booze in the liquor cabinet got more watery as the summer wore on. I lay with my head between speakers on the living room floor, listening to *Magical Mystery Tour.* But mostly I was wandering the shore of an inland sea. Stranded there, waiting, gathering strength to go back for another voyage, another year at Tabor. I didn't entertain the thought of giving up, staying home. I was halfway through and I had to finish it.

Yet I had learned something fundamental, along with advanced algebra, American history, and the periodic table. I had learned that my brain was accessible, mutable. I could change its chemistry, its balance. I could knock it out of one orbit and into another, if I tried, if I found the right substance. It wasn't something I thought about consciously, but it coalesced somewhere beneath the surface, colouring my sense of the possible. A safety valve, an escape hatch, a way out.

GIVING UP CONTROL

I walked toward the only pharmacy in town, about to get high on cough medicine. I was nervous. Cough medicine was a serious drug, wasn't it? Not an illegal drug, but definitely a bad sort of drug: it was supposed to get you completely wasted if you took a whole bottle, if you took the right kind. The fall days were rapidly shrinking, leaving more room for night. The wind warned of a bitter November soon to come, itself a prelude to months of frozen monotony. Winter would be long and empty, its only redeeming feature that it marked the passage of the year, the slow countdown to June and a final goodbye to Tabor Academy.

I was in a state of depression as dense as the year before. But the shock of it was long past. Now it was a familiar daily companion, a featureless fog laced with shards of anxiety, condensing often into self-contempt. I hated my own helplessness, my inability to control my life. But there seemed nothing more I could do. I divided the day into little survivable passages that I could take one by one. I just have to get through Math and then comes lunch. I just have to get through lunch and then I get a half-hour to wander by the shore. I steered my mood carefully to remain in smooth water, avoiding the rocks. Going with the flow. I developed a rich fantasy life that I could dial up as needed. I was a brilliant and powerful scientist-adventurer with a black belt in several martial arts. I had found a way to avert nuclear war by dominating the military leadership of both superpowers, West and East, interrupting their access to their own missiles by remote control from my island fortress. My beautiful girlfriend clung to my side during my television broadcasts, interrupting regular programming around the world—another electronic coup. I returned to this daydream and advanced it frame by frame, slowly, so as to savour it, whenever I had a few minutes to myself, walking from one building to another between classes or waiting for the dining room to open.

The night we'd spent getting drunk the previous winter had spawned a sort of friendship between Tennant and me. We'd got together for a few weeks that summer. We drove to Expo 67 in the Karmann Ghia his dad had bought him. And then we requested each other for roommates. But now Tennant was changing. In just a few short weeks, he'd begun to mutate from friend to foe. He didn't talk to me much anymore. He'd grown sly and distant. His smiles were smug, even mocking. I would come home to our room at Raven House and he'd be talking to Peter Smits, who roomed with Burdell, a short corridor away. I knew that Smits didn't like me. He had made a couple of remarks about Jews, so subtle you couldn't be sure they were meant to be malicious. "Do you like bacon and eggs, Lewis?" he'd say, turning toward me as I walked through, on the way to the bathroom. "I love bacon and eggs! But you're not allowed to eat them, are you?" With mock concern. Smits's style of nastiness was new to me. It wasn't like Moore's haphazard assaults, or Wiley's stinging one-liners. Its softness hid something darker, some inexplicable malevolence. He reminded me of a cinematic Nazi preparing for the torture scene, all smiley and sweet as he bustles about getting ready. In fact I was pretty certain that he and Tennant were part of some Christian brotherhood, and I was on the hit list. I couldn't believe that Tennant had turned into my enemy, just like that. Further evidence that the world was not only dangerous but incomprehensible.

As the October days grew colder, my fears began to come true. One night, I got home from prayers after dinner, as usual, to study, and found my ink blotter crisscrossed with anti-Jewish graffiti. This was an invasion, a desecration of my nice neat blotter: stars of David with "Oy" and "Jew" scrawled in their centres. Tennant was right there, at the desk on the far side of the room.

"Who did this?"

He shrugged stupidly. "I don't know."

"What do you mean you don't know? You were here."

"I wasn't here all day. And I didn't see anything."

"Did Smits do it?"

That shrug again. "No idea." He tried to smile and shrug at the same time.

A few days later, my drawers were turned upside down. Again there was no sign of any instigator. Just inference—and despair. The following week, there was some gooey substance on my sheets. I didn't want to guess what it was, but it really got to me. My private lair had been violated. How could I stop this? I confronted Smits once. He smiled sweetly, sympathetically. He would keep his eyes open. We would catch the guy, whoever it was. Next, I thought there were particles of something in my bottle of mouthwash. Every teenage boy had a bottle of mouthwash in those days. But I wasn't sure—maybe I was imagining it. There was no point in confronting anybody.

Acts of sabotage continued sporadically until February, when Tennant and I finally fought it out with fists, fingernails, and teeth. But that was a long way off. For now I was defenceless. I began to recede into suspended animation, my emotions a million miles away, with only the occasional, momentary fantasy of revenge. Anger would spark up and then disappear. It would ignite like a cheap Roman candle and then sputter and die. There was no specific enemy. Tennant and Smits were too much part of everything else. I didn't know whom to hate, or how to hate. Except that I hated myself for getting sent here in the first place. And then there was the emotion of shame—that constant companion of the victim who doesn't have the skill or the power to retaliate. Teenage girls become depressed because they feel unattractive. Teenage boys become depressed because they feel weak. And shame creates a centrepiece for depression. Shame is the dark chandelier that drains away the light. After the attacks began, shame nestled in close, my pet parasite, sucking away my sense of worth. There was just endurance, patterned by daily rituals of self-preservation. Fantasies to fill the day's monotony, hours spent in the library to avoid going home to my room at night. And then, when the library closed, the metallic taste of panic.

But this day was different. On this particular afternoon, I told myself I had a plan. Just saying it over and over banished the pooling depression and lengthened my stride. I felt alive.

"One bottle of Romilar extra-strength, please."

Did the pharmacist *know*? He must know. He looked down at me from his perch behind the counter and I felt the sweat spread across my face. But the thin line of his lips revealed nothing. Maybe he didn't care. And I didn't care once I was out the door. I walked firmly down the sidewalk, away from the pharmacy, holding my little treasure—a slim rectangular bottle of dark red liquid—in a hand wedged in my coat pocket. Nobody could see the bottle in my pocket or the thoughts in my head. Nobody could stop me.

Burton said to take one whole bottle, try not to taste it, and be sure you don't have to be anywhere for the next six or eight hours. Well, if I stayed out till after eleven I would have to answer for it, but not till tomorrow, and I might get home after Tennant and Smits had gone to bed, so I wouldn't have to face them. My mind was made up.

I was so naive. I had no clear sense of what "drugs" were. I wasn't sure whether this bottle of Romilar *was* a drug, like pot or acid. There was a lot of debate about pot and acid in magazines. But what was Romilar? It sounded like an ancient kingdom. Would this dark elixir take me to some faraway place? Would it take me into another land? Would it be hard to come back? Walking along in the slanted sunlight, I felt bouncy, light, and free. I was already half-high though I hadn't yet cracked the seal. Just by virtue of acting out.

So what's the plan? I could go to the library till it closed and then sit on the dock till everyone went to bed. It was hard to plan for an unknown event. Weak sunlight filtered down to the streets of Marion, Massachusetts, as I walked away from the downtown area—which really just consisted of a handful of stores. Past the impressive New England houses with their brown shingled roofs and pastel siding, sombre dormers balanced against stately profiles. Churches, lots of churches. Marion was a God-fearing community. But there were few people on the sidewalk as I approached the edge of town. I took the bottle out of my pocket, unscrewed the top, and smelled it. That was a mistake. My resolve stiffened and I up-ended it. Four ounces of very bad-tasting liquid, gone in as many seconds.

Now I feel naughty, a little bit dirty. I am walking around with no particular aim, except to spot a garbage can in which to dump the evidence. There is brisk, intermittent wind. The odd person passes me. We look away from each other as if there were suddenly some compelling sight elsewhere. I cut away from the main street, through sports fields and other open spaces, back toward the library. I feel a kind of peaceful emptiness, veined with excitement. I have taken a drug. What will happen to me?

Forty-five minutes later, I'm beginning to think that nothing will happen. I'm sitting on a bench outside the library building and it's starting to get cold. What should I do?

The thought of giving up and going to dinner is repellent. Four hundred boys sitting at twenty tables, each with a master and his wife in their places of honour. Bowing our heads, saying our prayers. Lunging into the soggy food while trying to keep an insipid conversation from dying entirely. No way. But now I think I feel something. No, probably just my imagination. Just the cold. Wait, what's that? I do feel something different. I'm not sure what. Lethargy creeping up my legs. Maybe it *is* the cold. But something else now. A hollow feeling seems to spread from my stomach. My body seems to be going numb. I feel distracted. My thoughts are skittering field mice. This stuff is doing something to me.

I get up off the bench and now the weight in my legs is becoming an impediment. It's an effort to walk without staggering, like being drunk. Big deal. Is this it? Got to get warm, but the library seems claustrophobic, so I walk away, down the path to Spring Street, which bisects the campus and curves away from the main road toward the forest. The sound of a car approaching catches my attention because of its disproportionate volume, its impossibly slow rise time, the climax of its roar as it passes by me, the lingering of sound filling my senses long after the lights have disappeared. I believe that the driver of that car knew me somehow. His majestic drive-by was a message. And then I notice that I'm thinking that being passed by a car was this epic event, this incredible performance, put on for me. How ridiculous. But I'm smiling. And noticing that I'm smiling makes me smile more. Something special is happening. And I discover an odd sense of accomplishment: I did it. I changed my brain. Now my skin starts to itch and there is a sensation of pressure in my head. I notice what must be my pulse pounding behind my face. I feel like my head is expanding, but also thick, full of cotton. I am staggering a little, but still pressing onward. My vision is blurring. The streetlights cast double images. I'm a little frightened: What will happen next? Could I die? What is happening to my brain? What is going on in there?

The oral ingestion of 400 milligrams of dextromethorphan hydrobromide has a massive impact on the nervous system. This drug is classed as a dissociative or a sedative-hypnotic. Imagine being hypnotized. You're no longer in control. You are not making decisions. Your consciousness is narrowed to an alleyway along which you coast peacefully as it stretches away in front of you. Flashes of perception go by like clumps of scenery on either side, while you float along with the slow, irresistible momentum of a dream. No more ruminating about what you should do, what you should have done. Just follow. The night is rich and mysterious. I'm entering the primeval forest, along a road prepared for me. It's like being drunk, but in colour, like when Dorothy got bonked on the head and transported to Oz, and there's company around here somewhere. The sense of familiar others. I am not alone.

Dextromethorphan mutes the cough reflex by slowing down the glutamate traffic. Like alcohol—but its impact is far more specialized. For one thing, it suppresses the normal functions of the brain stem (the stalk of tissue at the core of the brain) and cerebellum (the cauliflower-shaped hunk of layered cortex behind it) that coordinate muscular activity. That explains the staggering and the double images. But it takes special aim at a specific receptor site on the dendrites— the input zone—of neurons all over the brain. Dextromethorphan, and its cousin, the party drug ketamine, block the NMDA receptor, one of the main doorways for glutamate to enter a brain cell and help it to fire. NMDA receptors are among the most important gateways in the brain: they are the fundamental agents of synaptic plasticity, which really just means the capacity to learn. They are valves that activate tiny work crews of proteins—proteins that actually streamline incoming synaptic traffic—when something is important enough to be recorded, not just for the moment, but for days, weeks, months, the rest of your life. And because they're in charge of learning, they have to be smart. They have to discern what's worth holding onto. The trick they use to make that decision is called *coincidence detection*— they open their doors only when the sending and receiving neurons are in sympathy, resonating together, tuned to the same channel, coincident. In other words, NMDA receptors allow bonds to form among neurons that are already communicating. When sender and receiver neurons are in synch, the NMDA receptor opens its gate and swallows an extra helping of glutamate. That's when a casual fling between neuron A and neuron B becomes a marriage. They tie the knot, and that knot binds atoms of information into *sense*. "Perry looks pretty artistic with that scarf, just like Joe said . . . the guy's a maestro on the harmonica, goes with that devil-may-care style he projects." It's the coincidence between the pieces, and the harmony among the neurons that stand for those pieces, that forms a coherent image in the cortex. Fleeting associations get assembled into a model, impressions get turned into sense. NMDA receptors allow networks of neurons to shift, quickly and flexibly, to match the subtleties of what we blithely call reality—the shifting features of the world in relation to our own aims.

When dextromethorphan blocks NMDA receptors throughout the brain, that sense begins to fall apart. Reality stops getting through to the brain. And that's the essence of dissociation. That's why ketamine has been used as a battlefield anaesthetic. That's why, in dozens of experiments, ketamine has been used to mimic schizophrenia. And that means that I, and many millions of other teenagers over the following decades, figured out a way to go temporarily insane. But how does the cortex keep track of reality anyway? And what happens when it loses the thread? The cortex was designed by evolution to pick up, translate, and record a fantastic amount of detail about the outside world, so that we can achieve some degree of knowledge of the complexity of our surroundings. Lizards, with their almost nonexistent cortices, live in a simple world: hot, cold, movement, hunger, danger, escape . . . oh, and sex, even for lizards. But for us humans, the world is incredibly textured, and that permits us to tune our actions in accord with our knowledge, to act with purpose, with all the high-wire precision made possible by our intelligence. All we require for this astounding task is sense. And one other ingredient . . .

Sense is compiled by our NMDA receptors to produce a detailed and coherent map of reality, but we still require a motive to go there. That motive is what we might call *meaning*. Meaning is that special, personal insight of how the world is connected to *us*. But meaning doesn't just appear out of nowhere. We have to grow it: we have to learn who we are and what it is we want, what we need, what each change in reality implies for us—for our goals, our appetites, our fears and desires. We learn meaning from the earliest weeks and months of life—the goodness and safety of our mother's arms—until it starts to stabilize in adolescence and early adulthood, into that classic quest for love, admiration, and success. Meaning equals importance. It confers the royal kiss of significance on the outpouring of sense manufactured by the cortex. And it grows and consolidates in a nearby part of the brain: not the cortex, but the limbic system—a group of organs lying just inside the cortical shell. The main players of the limbic system are a bean-shaped tube called the hippocampus, responsible for the memory of facts and events, time and place; and just ahead of it the amygdala, which records the emotional colour and intensity of things experienced and antici-pated (see Figure 1). These limbic structures not only work together; they also connect to other neural organs that collect feelings from the body—muscle tension, shallow breathing, sweaty palms, and the tingling of terror from inside one's belly—so they know what *feels* important. And feeling is a fundamental component of meaning.

The limbic structures come in twos, right and left, with the cere-bral cortex wrapped around each of them like a wrinkled blanket of incredibly fine wiring. But here's the thing: there must be constant communication between these limbic structures and the cortex, between meaning and sense—in both directions. The cortex needs to feed the limbic system with sensibly organized detail—this is how

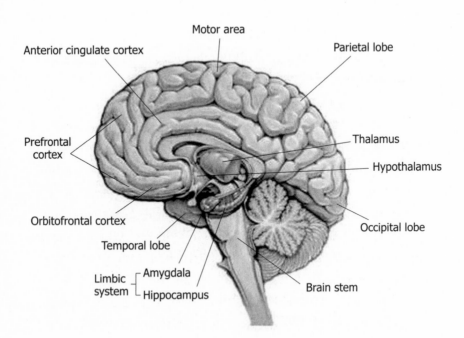

FIGURE 1. A cutaway picture of the brain, showing the five lobes of the cortex, but also the limbic system, thalamus, hypothalamus, and brain stem.

the world looks and sounds, this is what's going on and here's what can be done about it. Meanwhile, the limbic system has to charge the cortex with meaning: this is what I expect, what I want, what I need! This is what I remember. This is what's important. Nothing you do has any purpose without my prescriptions, and those prescriptions are simply . . . the past—everything you ever noticed or accomplished— distilled into a familiar stream of events, intentions, fears, and wishes. A well-functioning brain synchronizes limbic meaning, made up of feeling and familiarity, with cortical sense, our best approximation of reality, in a single, seamless exchange. That exchange is the cortico-limbic traffic that flows in both directions all day and all night, like the lanes of the Brooklyn Bridge.

NMDA receptors are the traffic cops, the toll collectors, that guide cortical traffic onto the bridge based on its sensibleness. They provide a constantly updated, fluid, and precise composite map of the world and one's place in it, and they send version after version of that map to the limbic system, where it is made meaningful. Then traffic from the limbic system returns to the cortex, to fine-tune vision (in the occipital lobe) and hearing (in the temporal lobe), but also to guide the tongue, arms, and legs (through the motor area), to act, to alter the world in the most beneficial way. The problem is that the NMDA receptors in my brain are now clogged with dextro-methorphan molecules! The glutamate isn't getting through. The receptor neurons aren't firing, or they're not firing fast enough. They just sit there, like dullards. And as a result, the pieces of the puzzle, the synchronized neighbourhoods of neurons, and the parts of reality they represent, come unstuck. Like a city in the grip of a plague or a riot. Communication is blocked, shut down. The cops are on strike. What finds its way onto the cortico-limbic bridge is a motley mob of broken down junk wagons, carrying a hodgepodge of impressions, but very little sense. Not only that, but this river of junk blocks the streets leading to the bridge, so bridge traffic wavers between chaos and gridlock. The routes between the limbic generators of meaning and the cognitive map of the real world are under construction. But meaning is still very much alive. Perry is beautiful. He reminds me of God. Meaning still sings its chorus of familiarity and attraction in my limbic system. The orchestra has never seemed as resonant, as pure, as it does tonight. But the reason it's so resonant and pure is that it isn't constrained by cortical input. Without an infusion of sense, the limbic chorus can sing anything it wants. The Brooklyn Bridge is down for repairs, and the result is *dissociation*.

You may wonder how this neural meltdown could possibly be appealing. But DM—that's code for dextromethorphan—and its cousin ketamine are among the most common drugs of abuse, especially for young people. At present, the use of ketamine as a party drug (or rave drug) is highest among middle-class kids who are typically in the eighteen-to-twenty-five age range. It is replacing Ecstasy as a social lubricant all over the Western world. Ketamine is also illegal in most countries. So why take it? One way or another, whether they are junkies or executives, people take drugs because they're not feeling right. The whole point of taking drugs is to change the way you feel. Drugs like DM, ketamine, PCP, angel dust, and those most damaging of substances, glue and gasoline, are called dissociatives because they do exactly what drugs are supposed to do: they dissociate feelings from reality, meaning from sense—and that's *all* they do. They don't speed you up, they don't slow you down, and they're not physically addictive. They just close down the cortico-limbic bridge so that your limbic system—the centre of meaning in your brain—is no longer tied to the world. Which is extremely convenient when the world isn't a nice place to be. And as a depressed teenager, I didn't find the world a nice place at all.

Meaning is supposed to go hand in hand with reality: that's the only way our brains could win in the gruelling marathon of natural selection. But when the world of meaning comes loose, when the limbic system breaks free of the cortex like a giant iceberg detaching from the mainland, dissociative drugs give the user an unusual kind of freedom. Disconnection from reality means: make it up. Take all the memories and associations, wishes and fears, that have formed and settled over the years of your life, and put them together however you like. Or however *they* like. Because those fragments of self have their own chemistry, their own hooks and latches. Once the dominion of the cortex is broken, those fragments will assemble themselves with shocking creativity. They will come together to form a story of swirling magic or whispering ghosts. And that story will unfold on the stage set of the world around you, based on the mash of sensory material that continues to occupy your cortex, but it won't be *of* that world. It will be entirely personal, entirely yours.

Now, with the sound of the passing car finally fading—now, when I just might want to be in control for a moment, to decide what to do and where to go, the dextromethorphan oozing from my bloodstream is shutting down NMDA receptors all over my brain, pressing giant coloured walls of untamed meaning in front of me, blocking my path. Like a crazed, leaderless army, these cough medicine molecules are taking charge, so that the pattern and beauty in the world around me no longer make any sense. I am staggering into the forest next to the road. The mat of leaves beneath my feet takes on a bizarre, fuzzy texture, shocking because it looks like fabric, not leaves at all, as though the forest were covered in broadloom. It emits its own light, an iridescent green that doesn't belong in the dark of the forest. A clump of matter at the base of a nearby tree suggests a massive, living organ, seeming to pulse with its own heartbeat. Nothing makes sense.

39

But my consolation prize is an extravagant stream of impressions . . . and a compelling familiarity. Limbic meaning makes everything seem familiar—the sinuous branches, the whitewashed shoulder of the road, the gleam of wetness on the grass. My ears roar with the sound of reality draining into a dark sea, a cliff eroding with the pounding of the waves. No, it's my heart, or a gang of African drummers massed behind hillsides of thick orange light. I drift farther and farther from the road, deeper into the forest. I have no idea where my feet are taking me. And I hardly care. The grey air is pregnant with meaning. It is so thick that it seems to merge all the colours and all the sounds of the forest. There is too much light for nighttime. The air quivers, clustering in waves. I momentarily see a spongy beach ball, which then turns into a rubbery mess of rotting leaves. I am lost in the corridors of my limbic system, where cartoon images bubble from the pastiche of my past. And now the familiarity overwhelms me, condenses into the background hubbub of almost distinct voices. A crowd of pedestrians on a busy sidewalk. But I can't see them. I can't really hear them. It's as if they're just out of earshot, more like thoughts than actual voices. The impression of other souls is irresistible, immensely comforting.

I make my way along a path that forms itself just ahead of me, undulating through the woods. My legs are heavy and my head hot and huge. The overhanging trees are not only familiar, they're compassionate. And now that I notice, the embroidered patterning of their branches is impossibly beautiful. They reach out to welcome me. I feel that they have been waiting for me, that these patterns are offerings of beauty, bequeathed to me, to bind my wounds, to heal me. When I close my eyes and just stand still, the sense of presences wells up from inside: background whispers again blossom into voices, saying outlandish, made-up things, like a reverie turned up to the volume of a waterfall. I open my eyes and look around once more. I feel like a new person in a new world, drenched with the warm rain of total intimacy.

Several hours later, I am back at the rocky shore of the sound, sitting on the dock behind Raven House. The waves roll past me with biblical significance, as though I am Noah, the first sailor, the only sailor on the only boat in the only sea. And God is somewhere around here: my great and good hypnotist, looking down kindly on me. My feet dangle over the edge of the dock, and the waves entrance me with their rhythmic persistence. I'm not sure how I got here from the woods along Spring Street. I feel like I've surfaced at the end of a long dream but I'm not awake yet, not wanting to be. The pressure in my head has diminished. The few moments of fear are long gone, and I feel truly happy. I am sitting fifty metres away from my dormitory room, where my tormentors must now be sleeping. I am hours away from another joyless dawn. I am stuck in October, which is nowhere close to Christmas, let alone June. I'm as bricked up in my prison as I could ever be. And yet I've escaped. I've found a new world. And I know how to get back here whenever I want. I'm triumphant.

I'm also in big trouble. Mr. Sukert lurks at the front door of Raven House like a patriarchal behemoth, very much awake and waiting for me to show up. He asks me where I've been.

"I don't know," I say, realizing abstractly how wrong that sounds. My voice is distorted and thick. The neural impulses responsible for co-ordinating the muscles of my jaw and tongue are still a sloppy mess. The *d* is difficult to pronounce, and I have to work at it two or three times.

"What do you mean you don't know? Do you realize what time it is?"

I try to reply, but poor articulation and an utter lack of anything to say weigh against me. I manage only "no," a nice soft syllable, easy on the tongue.

"Well it's two-thirty in the morning, and I have been waiting up—are you all right?"

"Uh-huh."

"Are you drunk?"

"Uh-uh."

"Where have you been?" Mr. Sukert can't really get angry. I begin to like him a lot.

"Mr. Suk-sukert, did you know . . ." Dormant brain systems begin to function and I stop myself. "I'm sorry. I'll go to bed now."

I was not allowed out in the evening for the next two weeks, and that was hell. But nothing could quite take away my private sense of accomplishment.

3
INTO THE FIRE

It was November. I'd begun to feel a strange sense of dislocation, especially in the morning and then again at night. I would sit in morning assembly while Witherstein droned on and on, trying to dissociate through an effort of will. I tried to make myself go a little crazy. "You boys are well aware of the rules concerning visits to the restroom during study hall." What would it be like to be nuts? To stop caring entirely? "Congressman Hiller will be with us for the weekend, Tabor being his alma matter." Congressman Hitler. Doesn't he mean Congressman Hitler? "I feel nothing but pride when I look at you boys." If I narrow my attention to the spaces between events, can I make reality disappear? I stare at the back of the chair in front of me until my vision blurs and sounds converge into a roar. I am mimicking the first hour of a DM high. "A special vespers service will be held before dinner on Sunday. Formal attire is expected of everyone, and that doesn't mean corduroy, does it, Mr. Stillwell?" A mighty fortress is our God. Swelling voices in Cartoonland. Dear God, please let me dissociate.

For weeks I practised going blank. I held onto that blankness on the way to class and returned to it at mealtimes. My daydream of world domination faded like the radio signal from a town long past.

At night I roamed the streets, looking for anything that broke from the familiar. I walked along the rocks lining the shore, feet freezing and wet. I went into an unlocked church and sat in a pew halfway back, in almost total blackness, pretending to worship this generic Protestant god. I walked slowly along the empty streets, imagining that I was the only person left on earth.

But Tabor would not go away. French class was dominated by four or five enormous louts. Football players. Their legs stretched out a row ahead of their slumped bodies. They lolled in their seats like giant reptiles, deriving a lazy satisfaction from Mr. Barker's rage. I feared them because of their power and their indifference. They would not pronounce these French words properly. Mr. Barker could not make them do it. Their smiles, shrugs, and yawns defeated him repeatedly. He would end up insulting them, his face turning red and his crooked teeth protruding helplessly. They reduced his life to a mockery of teaching. Yet he was more fortunate than Mr. Peters, my English teacher. Mr. Peters had one stunted arm that waggled like a fish when he wasn't careful to pin it by his side. I worshipped Mr. Peters. He showed us how to look beneath the surface of narrative to discover layers of meaning in Conrad, Lawrence, and Melville. I thought this was a wonderful trick. It filled the empty places of the world with vision and intelligence. But others referred to him as "The Claw," and this horrified me. He would hear. He *must* hear. They were not careful to make sure that he was out of earshot. They were capable of anything. But some masters were different. They belonged at Tabor. Their steely eyes narrowed to gun barrels when they detected opposition. One day, Mr. Tudge dragged me out of Math class by my shirt collar for speaking to the boy beside me. I hadn't meant to be disrespectful. I forgot myself, sir. But most of the teachers were sad men of all ages, with little in common but their exile. They were lower on the totem pole than the coaches, and that must have gnawed at them.

44

My friends in Pond House had all graduated last spring. Burton was cutting trees somewhere. Perry had gone to a big Midwestern university. And Norris was gone too—I didn't know where. Norris had disappeared from the daily rounds of meals and classes in late May. Given his unhealthy complexion, spindly body, and cobra-like tongue, nobody seemed eager to dig him up. Finally, Decker, the assistant headmaster, the man we feared most, went to his room one morning. According to legend, Decker found him lying in bed reading Nietzsche, and, like Nietzsche, he'd gone mad. Schwartz said there was no way they'd publicize Norris's insanity in the last three weeks of school. Better to let him slink off to whatever pathetic future awaited him. Norris had won, and in winning he imparted a golden lesson: when all else fails, go nuts. Then no one can touch you.

Now Schwartz and I were the only ones left. We saw ourselves as the dregs of a tiny but exalted civilization who had scattered before the marauding hoards. This year, he and I were the seniors, and it would be our turn to leave when June next came around.

Thomas Gelsthorpe was the one Pond House graduate who kept in touch and who would soon become my guide to a different world. He had gone to Berkeley. Yes, Berkeley, that mecca of hippiedom. It was mid-November when he first wrote to Schwartz and me—one long letter with no capitals or periods, minimal punctuation, a stream-of-consciousness poem glorifying the West Coast. I pictured his pudgy features and scholarly glasses as Joe read aloud: "The sun glides down giddily on girl after girl after girl illuminating their loveliness their breasts their bodies their breasts breasts breasts the girls wear no underwear where you ask here here in california HEAR HEAR no underwear and everybody smokes grass everybody all the time smokes grass and drops acid you poor miserable mildews i wish you were here you must come!"

We wish we there, too, Thomas. Oh do we ever.

———

I visited Schwartz's lair often. He would lie sideways on his bed like the caterpillar in *Alice in Wonderland*. I would sit on his desk chair. There was no other place, no other furniture, and I would lean back and listen to Joe Schwartz, philosopher, waxing eloquent on the mysterious forces that had brought us here. This is society's crucible, he explained, where the next generation of military-industrial leaders is being forged. When all human sensibilities are neutralized, we'll be allowed to go home. But by then it will be too late. Schwartz was the master, the poet laureate of pessimism in the grand style of adolescent eloquence. And yet he had a reserve of indifference and self-confidence that I could only envy. He lived in Puerto Rico, where his parents were university professors. This seemed fantastic to me. He spun scenarios of an island paradise where we could be free to drink rum and ride motorcycles, smoke marijuana, and chase women. I didn't know what it would be like to chase women. It sounded scary if you thought about it, but Joe would know what to do.

He put down Gelsthorpe's letter and looked at me intently. "We have to go," he said.

"Sure, but how are we going to do that?"

"I have a plan. I think my plan can work."

"What plan? What do you mean?"

"Listen, Marc. Let's be a bit creative," he explained patiently. "We could apply to complete high school in Puerto Rico. I know the school. It's near the university where my parents teach. We could stay at my parents'. They're away half the year anyway. They won't mind us living there. We can go to school, which ends early in the day, and then ride around the island, find beaches. Beautiful beaches. We could do that. It's not impossible."

"Schwartz, you are truly nuts. They're just going to let us start in the middle of the year?"

"We could have our credits transferred. People do. Christmas break is the ideal time to switch schools . . ." He looked at me with mild disapproval, as though I should have thought of this myself.

Schwartz and I began to construct a dream that progressed rapidly with the passing weeks. We got together often to scheme. He was in charge of correspondence with schools and parents. My role was that of sorcerer's apprentice, bringing up details that Joe would fit expertly into place. My depression subsided partway. I stopped trying to go blank and began to sense a future with some sweetness in it. I called my parents. I said I was doing well. I always said that on the phone. I practised sounding well for several minutes before calling. I told them I had a surprise for them. My mother was curious about it—a little suspicious. My father said good luck in his usual way. I reported back to Schwartz: I'm getting them ready. I think they're going to go for it. Schwartz nodded sagely.

Three weeks later, I arrived at his room in my now-familiar state of excitement tinged with dread. He looked up at me with a dark little smile.

"What's up?" I asked.

"We have to get there sooner, not later . . ."

"What do you mean?"

"I haven't heard back from the school yet. Christmas is in two weeks. My parents are in Asia somewhere. The only way to get out of here is to do it ourselves." He spoke quietly, sounding much more grown-up than I felt. "Can you come down for Christmas holidays? It's just possible that we'll end up staying," he ended sardonically.

"I have to go home! I have to see my folks . . ."

"So? We'll meet in New York, after your visit home, only instead of going back to Tabor we'll just keep going . . . to San Juan. If that's all right with you." He grinned.

———

47

I flew to Toronto as soon as school let out. I rehearsed my lines, and more importantly my tone, in the garage before dinner: I was a solid, confident, mature sixteen-year-old—a cutout from *Young Hobbyist*. I announced my plan to switch schools. But after twenty minutes of strained improvisation, I began to see that they weren't really listening.

"Mom, it's almost all set up . . ." I insisted, beating back the momentary threat of a sob.

"What exactly is set up, Marc?"

"The school. It's supposed to be a really good school, especially for international kids. It's very oriented to the arts and politics and stuff."

"You still haven't told us the name of the school," she remarked, and I tried again to remember it, without success. I wasn't sure Schwartz had ever told me. I twisted away from her eyes. The gleam of Puerto Rico began to flicker. She turned up the frankness dial. "We can't take the chance of you losing your year, Marc. It's too late to switch schools this year. Don't you agree? Just a few more months and you'll be leaving Tabor behind."

Had it ever been real? Could Schwartz somehow resurrect it? I wanted to get on a plane that night, but I had five more days slated for Toronto. Images of another Tabor winter invaded my reveries. Depression churned in my stomach below bright, flitting thoughts of escape. I wasn't ready for this. The sense of hopelessness came on so fast it panicked me.

So I bought a bag of marijuana from a grade-school friend.

I had tried pot for the first time three or four weeks earlier. In the cold wind of early December I had huddled with Schwartz and another boy and sucked acrid smoke from three skinny joints, one after another. It wasn't until the last one was gone, and my throat was raw with coughing, that I began to feel an effect. People said it takes a lot the first time. But I was exhilarated . . . at the miraculous waterfall of thoughts and images, each perfectly formed, beautiful, detailed, and exquisite. Yet conjured up by me. Just by thinking a thought, I invented a new wing in a mansion of ramifying corridors. And I was happy. Once again, a drug had shifted me from a hopeless reality to a compelling new landscape, following a path that kept creating itself. Once again, here was a thing that I could take into my body and change the way it felt to be me. So I was already enamoured of the possibilities of pot before my Puerto Rican fantasy began its death throes. I needed it now more than ever.

I smoked up a few more times during my remaining days in Toronto. I could sit in my parents' home and feel less than humiliated by the stupidity of our dream, less than angry, no longer sad. I could change the way I felt *at will*. As long as I had my bag and my papers, I was safe from depression, free of the nausea of shame.

On a snowy night in late December, I met Schwartz at LaGuardia airport, as planned, but he didn't seem to have tickets for any flight to San Juan. I was disappointed. I felt I should be even more disappointed. When I told him of my parents' negative reaction, he just shrugged.

"Aren't you upset?" I asked.

"It was never going to work," he said. "And all the flights are cancelled for tonight, anyway." There were storm warnings on the radio. "We'll spend the night in Manhattan," he concluded, as if that had been his intention all along. I didn't get it. What was he thinking?

Within two hours we were at a Howard Johnson's restaurant somewhere near the Village. I gazed up and down the counter, mesmerized by the creatures of New York. The poor people, the street people, the blacks you never saw in Toronto or New England. The interior of the restaurant was warm and steamy, providing shelter from the snowstorm outside. I was lulled by the warmth. Over and over I told myself: *We're not going to Puerto Rico*. Not tonight. Not tomorrow. Not ever. And suddenly I wanted to try my grass. I wanted to make sure it was strong enough. Because, if it was, then I'd have my own remote controller, my little mood modulator, and what power that would give me! Power over the weight of despair that was coming, coming, with Tabor about to fold me up again.

Now I'm on the toilet seat in the downstairs restroom. I pull out one of the two joints I rolled before leaving. The other joint is nestled in the baggie with twenty-five dollars' worth of loose grass. I smoke slowly, gently, trying not to cough. Schwartz must be wondering what's taking me so long. The smoke drifts through me and warms me. I think about Puerto Rico. I think about home—my parents' indecipherable faces. The graffiti scrawled on the stall door attracts my attention. BEST BLOW JOBS and a phone number. What a strange thing that is. It sounds like an ad for something very grand: a Broadway musical. I can't take my eyes off it because it seems funny, though it appalls me. And then I realize that I am experiencing the world in a different way. My attention is curling around its objects. It lengthens, becomes a telescoping tunnel, where my path leaves tracings that so intrigue me. I can't help but look back as much as forward. Not a tunnel but a hallway with etchings on the walls. I examine the etchings—my thoughts—and find in each a labyrinth that unfolds further, leading to new galleries, where my thoughts form new etchings, each capturing my attention for many minutes.

I follow along and I keep losing myself in branching galleries. What are these rooms? And what was it in the last one that I can no longer remember? Everything has become so immediate, and yet the immediacy creates a maze of unmarked turns. BEST BLOW JOBS is there like a beacon. I'm smiling. But I'm also amazed: Why is every thought its own world? Where is the main artery of this jungle, the artery I call "me"?

———

How does the brain shift from a well-coordinated orchestra to this series of solo acts? As if one instrument has declared its independence from the conductor, followed its own score, invented its own score, only to be replaced by another prima donna? Does this profound but comedic high, experienced by people all over the world, spring from some kind of damage? I wonder: Have I poisoned myself? Or have I released myself from the rein of an internal dictator?

The drugs I've tried before, alcohol and dextromethorphan, have a massive impact on the action of the major neurotransmitters, glutamate and GABA. But the molecules in those bitter liquids look nothing like the molecules in our bodies. Their impact comes from brute force, not from mimicry. Cannabis is different. Like most of the illegal drugs I have yet to try, cannabis contains molecules— cannabinoids—that resemble molecules produced in our very own brains. Cannabinoids are natural brain chemicals. They have a purpose. But that doesn't mean that our brains are designed for an endless pot party. Our self-made cannabinoids circulate in much smaller quantities than the torrents of molecules sucked from my joint. It's the same story with a lot of other drugs, including opiates and amphetamines. Most of the drugs that get us high activate chemical messengers already present in our brains, or act just like them, but they come in quantities—or with exaggerated properties—that evolution never intended.

Cannabinoids are specialized neurotransmitters released by neurons that have only just fired. Normally, neurons take a little break after firing. They become unresponsive for a few milliseconds. That's the brain's way of preventing the more vociferous neurons from taking over the party, running off at the mouth, going on and on after they've already spilled their little message to the next neuron in line. But cannabinoids interfere with this neural etiquette. They declare "More of the same!" At least in some parts of the brain, cannabinoids increase the firing rate of the neuron that has just released them, through their influence on the neuron just before it in line. Imagine that you are a polite conversationalist, and you usually wait for someone else to speak after you've had your say. That makes conversations sane, balanced, and stable. Now you've just taken a magic brew that urges you to keep on talking, so nobody else gets a chance to speak. By increasing the action of neurons *that are already active*, cannabinoids cause each thought, each response, each act of perception or imagination to magnify itself. And let the other voices be damned! This is a form of neural plasticity—the tendency for neurons to increase their responsiveness. It is almost the opposite of DM's shutdown mechanism. More like the normal function of NMDA receptors, it sensitizes active neurons. It allots plasticity to neurons that already have a voice, increasing their influence, but ignoring the larger etiquette of cooperation.

Cannabinoids enhance with exponential power whatever show is playing at the moment. Thoughts and perceptions are self-amplifying. That's what produces the ever-ramifying corridors of mental exploration whenever you smoke pot. You don't see the big picture because you're so caught up in the momentum, the unstoppable thrust, of your present reflection. It's about this! This is what's important! This is the most important, compelling, significant, profound thing ever! But if the brain produces its own cannabinoids, albeit in limited quantities, then there must be a purpose for this kind of mental distortion. Evolution wouldn't waste hundreds of million years perfecting a neurotransmitter system whose main purpose was neuronal self-aggrandizement. Unless there was some benefit for the cognizing organism . . .

The cannabinoid receptor system matures most rapidly, not during childhood, not during adulthood, but during adolescence. So it wouldn't be surprising if cannabinoid activity is *meant* to be functional during adolescence, more functional than at any other period of the lifespan. As far as evolution is concerned, adolescents might well benefit from following their own grandiose thoughts, goals, and plans. By doing so, and by ignoring the weight of evidence — or sheer inertia — piled up against them, they would greatly amplify their tendency to explore, to try things, to imbue their plans with more confidence than they deserve. The evolutionary goals of adolescents are to become independent, to make new connections, and to find new territory, new social systems, and most of all new mates. The distortions of adolescent thinking might be precisely poised to facilitate those goals.

In fact, teenage thinking bears an uncanny resemblance to the delusional profundity of a marijuana high. Adolescents ignore most of what their parents think, most of conventional wisdom, and are completely spellbound by their own ideas. They follow chains of logic that nobody else finds logical, and voice excessive allegiance to their own predictions about how things will turn out. Even when they're not stoned, adolescents live in a world of ideation of their own making and follow trains of thought to extreme conclusions, despite overwhelming evidence that they're just plain wrong. In fact, this self-augmenting and highly personal delusion is just the kind of fantasy Schwartz and I had been concocting these last few months. We'd been blithely building our scenario about school in Puerto Rico, getting lost in its intricacies and ignoring its sheer unlikelihood. Until the bubble finally burst over Christmas vacation. You could say that Schwartz and I had created our own brand of dissociation. But it wasn't the DM-style knock-you-over-the-head departure from reality. We'd gotten lost in the possibilities of how the world *might* be, not in the dark matter that emerges when you leave the world behind. The audacity of our plan may have been bolstered by the maturation of cannabinoid receptors, fuelled by the flow of cannabinoids made in our own cells. And if cannabinoids help teenagers design their own reality, then maybe my new preoccupation with pot wasn't so much an aberration as a second, somewhat desperate effort to keep my options open.

But it's 1967, and pot is highly illegal. So here I am, sitting on the toilet in a Howard Johnson's restaurant, following some internal call of the wild, but with my teenage brain tilting and flowing a bit more than evolution had intended. And I don't realize how crazy this is. My neurons are unleashing torrents of self-mesmerizing information. Information that becomes its own purpose. Each thought takes on enormous weight. My senses are wide open, and yet I am propelled down one cognitive spillway after another, fascinated by each bend in the road of rumination. Until the next thought, the next neural rebellion, builds enough firepower to take the stage. And then a new corridor opens up in the labyrinth of possibilities, and down I go. And I can't remember what I was just thinking because the spillway I'm on is just so momentous there's no going back. I can't even remember I'm in a restaurant in New York. No way to reduce the forward thrust, to settle down and put things together. Puerto Rico . . . I smile more broadly. What was that about? I sit there grinning, with BEST BLOW JOBS shimmering in the background, except that now, incredibly, there is a terrific pounding somewhere inside the restroom. Something exciting and scary is going on. Someone is pounding on the wall. And I don't realize for quite a few seconds that it's the door of my stall that is shaking with the blows. Why would somebody do that?

"Open up, kid," comes the voice of a TV cop. I think for a fraction of a second that this is funny. But what if it's a real cop? And now my musings freeze into a brittle, horrible scenario. Reality lurches into view like the station at the end of a train trip. Have I slept through the whole thing? I am in New York. I am smoking marijuana. That's against the law! I stupidly stuff my bag of pot in the pocket of my coat. Why do the police want to arrest me? I'm just a kid. I start to get up at about the moment that the door crashes inward to reveal two bulky New York cops. One reaches in, almost gently, and lifts me the rest of the way to my feet. He ushers me out of the stall and stands with me while the other goes through the pockets of my coat, still hanging from its hook, and pulls out my treasure. I am alarmed at the prospect of losing it. I still don't understand that there are far worse consequences at hand.

Everybody turns and looks at me as the two policemen escort me back through the restaurant. Some are smiling, but Schwartz looks deadly. His expression is one I will never forget: a blend of horror and pity. My burger and fries are sitting there waiting for me, and I wish that I could just sit down and eat with Schwartz, but I see now that I am going somewhere entirely different. Not Puerto Rico, not a night on the town, but an unplanned journey with no return ticket.

I spent the first hour of my first bust in a New York City holding tank. Black men in pink pantsuits leered, flirted, and spat obscenities. White men, grey as corpses, leaned against the bars and gazed at nothing. Who were these people? I couldn't move or speak. Occasionally one looked at me with fishlike eyes, and my gaze wheeled away in terror. A row of men like a menagerie of beasts, each so unique, so distinct, yet so improbable. A fat, misshapen white man with scaly skin stared at me and edged closer. Every time I looked up, his gaze was waiting. Fear like a steady electric current bolted me to the floor. But then the same two cops appeared out of nowhere and led me away, through corridor after corridor, into the back of a police van. A short bumpy ride and they opened the door and ushered me through the corridors of a different building: Juvenile Hall, as I was soon told. I stopped to barf in the corner of a passage in this labyrinth of prisons, and they waited with surprising patience until I was done. They understood my fear better than I did. Finally, I was shown to a single cell, and I collapsed on the mattress and slept.

The next day, I was told that my father had arrived. His presence seemed heroic, magical. For a minute I imagined he had come to rescue me, not just from the police but from Tabor itself. But then I saw that this was another stupid fantasy. I sat in a small, locked, windowless waiting room, trying to make sense of it all. Was getting arrested equivalent to getting sick, or going crazy? Maybe it will work. Won't they have to bring me home now? I heard Schwartz saying, "Don't be naive." My father must be angry. He must be. My mother must be very, very disappointed. The door opened and I was escorted to a large foyer where parents, kids, and lawyers mingled in small groups. And there was my father. He hugged me quickly, gruffly. He asked if I was okay, and then we stopped speaking. Soon we were joined by a lawyer, and we shifted to a hallway echoing with male voices. I tried to get rid of the taste in my mouth at the drinking fountain, as well-dressed men argued jovially with each other. My father now seemed part of the club, visibly relaxed. But I couldn't look at him. And I didn't want him to look at me.

The judge consented to a six-month term of probation, and then there was much shaking of hands and exchanging of papers. I did not know whether to feel disappointed or relieved. Nor did my father. With our eyes on the passing stream of Manhattan storefronts, we sat side by side in a New York taxi and spoke in comic-strip balloons. He told me I was lucky—lucky that I could go back to school, not to jail. The distinction was temporarily lost on me.

"I guess I deserve to go to jail."

"Well, what you did was really stupid."

"I know. I'm really, really sorry."

"That doesn't help much."

I was abject, ashamed, cringing in my seat, wanting to drop into the crack between seat and door. Contemptible. That's what I was. Unbelievably stupid, unbelievably irresponsible: selfish, selfish, selfish! But that wasn't quite it. What described me, what this inner voice accused me of, wasn't exactly selfish, not exactly weak, but some meridian of self-blame that included both, and also dirty, disgusting . . . maybe just *bad*.

I berated myself with masochistic vigour, warming up for what was to be a month of masterful self-loathing. I had shamed my family, brought them nothing but expense and worry. I had shamed myself. The whole purpose of surviving Tabor was to complete the moulting process in relative darkness, a lizard in a cave, shedding the skin of childhood to emerge strong and free, independent, confident, manly. But here I was being driven by my dad to the airport, to return to school—while he went home. The sense of badness that had sifted and settled over these last two years now condensed into a purified essence. I was sentenced, convicted, and not metaphorically, as Schwartz and I liked to play it. I was really convicted. I was a convict.

Back at Tabor, I tried hard to be good, and I showed the imagined faces of my parents, my teachers, my probation officer, how good I could be. I sensed myself observed by a ghostly audience, and I purged myself of weakness, shunning marijuana when I came across it. But I was frozen inside, stunned by shame. So I tried to be still, not to move. I tried not to be noticed—not only by the others but also by myself. To be sufficiently stationary that nothing would activate the waiting images.

Yet other me's flickered into existence, gathering substance from coalescing particles of misery. I would somehow get them: my parents, my teachers. I would not walk obediently to the gallows. I would show them. Wait and see. And a third fragment materialized as well—not rebel but pragmatist. This me was scanning the horizon, looking for a way out, an unguarded exit. Ready to run away from my real and imagined wardens and set off on new adventures. Had I been born a hundred years earlier, I might have gone to sea. But the sea at Tabor was caked with ice and the boats were locked away. Instead, I looked to California—the opposite side of the world. There the drug movement was getting underway, nicely shaping up to receive new conscripts, and the girls—the girls who wear no underwear . . . Thomas Gelsthorpe left no doubt that the girls would be waiting.

My grades were astounding that long winter term. I stood second in a class of 120 seniors. I lifted weights. I waited tables. I crushed my depression into a lump of coal, and that coal into diamond. I read Conrad and Lawrence. I wrote poems about life and death. I dreamed about next year, about going to college. I applied to Berkeley, Stanford, and UCLA. I wanted to go and live with Thomas Gelsthorpe, to frolic with the girls. I wanted to try LSD. I would get on board the family wagon train, and I would get off when we reached the Pacific. There I would say goodbye to my parents and brother, and I would find my freedom.

I returned to Raven House at night, as late as possible, but attacks by Smits and Tennant recurred every few weeks. Then one day Tennant grabbed my leg as I climbed off my bunk to shut off the insipid tripe emitting from his stereo. That did it. A completely unfamiliar gush of rage shot through me. I snatched him by his shirtfront and smashed his head against the bedpost. He grabbed my neck and tried to force his fingernails into my face. We fought and fought and fought, neither of us expert enough to land a decisive blow. But the room was wrecked and both our faces were bleeding and bruised by the end of it. That was in February, and it put a stop to the attacks for good. I was left alone. I no longer felt like a passive victim. But the world seemed a dangerous place.

DOPAMINE AND DESIRE:
A ROMANTIC INTERLUDE

My last half-year at Tabor was complicated by a new problem, one that brought unexpected pleasure along with unavoidable pain. Every month, starting in late January, I had to invent an excuse for returning to New York to see my probation officer. "My grandmother is sick . . . again. I have to fly back to Toronto this weekend." Even my parents participated in this charade, underwriting my excuses through phone calls to the Assistant Head. "We can't have them know about your . . . situation. There's no point in that." Each time I went as far as New York, but there was no ticket out. I took the train in from LaGuardia, then the subway down to the Bowery. I got off and walked for half an hour through canyons of enormous grey buildings, past block after block of pathetic-looking people in dirty clothes. I had just turned seventeen, and depression tracked me with renewed vigour.

I'd go up to the fifth floor of a building that occupied an entire block, then down an overheated corridor of shabby offices. I'd knock on a door that looked like all the others, and my probation officer would open it with a lopsided smile. He had stains on his wide flowery tie, and I liked him because he seemed friendly and he wanted to know how I was doing.

"Not too bad." I melted at any sign of sympathy. I was that hungry.

"You gonna have some fun while you're in New York?"

"Yeah." Looking down.

"What kinda fun?"

"I don't know."

"Where you staying?"

"With my parents' friends in White Plains."

"White Plains! Nice place. What do you do there for fun?"

"I don't know. Maybe I'll go out with Lisa, their daughter. She's my age."

"You like her?"

"Yeah. She's really nice."

"Want to get a little action going, huh?"

"No. I don't know. It's probably not that kind of thing."

"Wouldn't mind, though?"

"I don't know. It'd be weird, since she's my parents' friends' kid."

"She pretty?"

"Yeah."

"Hey, having some fun with the girls, that's okay. Just stay away from the grass. Can you do that for me?"

To my amazement, Lisa and I got some action going the very first night. We went to an Italian restaurant in Manhattan. She led, I followed, gratefully. We drank wine for hours. We drank our faces off. Nobody cared how old we were. We went back to her parents' place, and they were asleep! They'd left a nice note. She pushed me down on the sofa, plopped down beside me, and started kissing me. Huge kisses, soaked with wine. I couldn't get enough of those kisses. They were nothing like the arboreal orgies with Betty the Machine, a girl who liked to meet Tabor boys in the woods at night. Instead of disgust, I was gripped by that most fundamental of sensations: desire. Which is a poetic word for *wanting*. In this case, wanting and getting. Wanting something I could have, and getting it. The touch of skin and tongues and the shared secret of being naughty, the gaze that synchronized our drunken brains and that promised some kind of bond we had never ever had with anyone else nor ever would again.

Back at Tabor, I began to believe I was in love. It had just taken one night. Wanting and getting and wanting and getting. And being bad. Nobody knew. It got me past the depression of my next New York sojourn. After my visit to the probation officer in February, Lisa and I spent the weekend drinking wine, and I told her how I had finally bested Tennant in the fight of the season. I described a scene of splintered furniture and pools of Tennantian blood. And then, as befits the victor, I was permitted to kiss her, and then fondle her breasts, among the cushions of her parents' sofa. She made life tolerable in February. She was the goal of my existence in March. We necked until our mouths ached. And after pulling my hands up firmly from her thighs, she would let them settle around her breasts. My hands around her breasts. Wanting and getting *breasts*. I was reborn.

Like most teenage boys, I had often imagined being with girls. But Lisa was more than a fantasy. I had access to her. Her breasts bounced palpably when I successfully unhooked them. Even at Tabor, she was not out of reach. Far away, but not out of reach. And I reached for her daily, with neural muscles that grew stronger with the isometric tension of wanting. Wanting, wanting, wanting. Buddhists call it craving, and they rightfully claim that it is the centre of selfhood, the foundation of personality. I craved Lisa more than I'd ever craved anything. And then, once a month, the climax of the show, the piece of cheese at the end of the maze: the getting.

———

The cortex has a back half and a front half, as shown in Figure 2. The back half is in charge of perception, and the sense it makes is the sense of how things look, sound, and feel—how things *are* in the world. But the front half, mostly made up of the prefrontal cortex, is in charge of action, and its job is to make sense of how things *can be*—how the world can be changed, transformed, through the exercise of the will. Its job is to plan, to formulate goals, to compare options, and to choose the option that is most likely to please. The back half of the cortex gets its meaning from the limbic system—the amygdala and hippocampus—as already described. But the front half, the prefrontal cortex, gets most of its meaning from another set

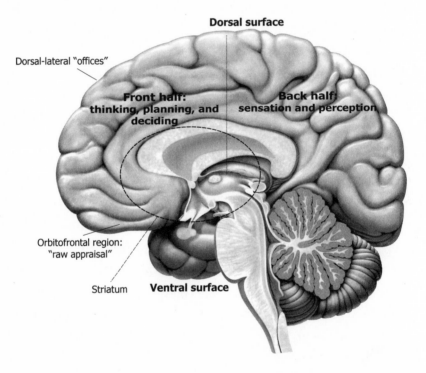

FIGURE 2. An image of the brain, split down the middle, highlighting two dimensions: front vs. back, and dorsal vs. ventral. The striatum does not show up on this image, but its location is marked by the oval.

of subcortical structures, very much like the limbic system and inter-laced with it: these structures, taken together, are called the *striatum*; they are nested around the base of the cortex and connected to it through their own set of bridges. The so-called reward systems of the brain are housed in the *ventral* (lower) striatum, and these rewards— the glowing anticipation of pleasure and success, the feeling of satis-faction when goals are attained—are what drive the frontal cortex to act on our behalf.

The prefrontal cortex, like most big companies, has different functions on different floors. On the higher floors, the *dorsal* heights, reside the executive offices. Many of these offices are lined up along the outsides of the dorsal prefrontal cortex (you can't see them in Figure 2, which shows the middle of the brain). In these offices, a precise and sensible model of recent events is created and re-created each moment, according to a capacity called *working memory*. The *memory* part is just the holding onto details—whatever events hap-pened in the last minute or so—and the *working* part is fitting them together so they make sense. A little farther back along the dorsal cor-ridors, closer to the middle, lies another suite of offices, and this is where decisions are made. Here, in a wing called the *anterior cingu-late cortex*, competing ideas and strategies are compared, rapidly and intelligently, and a corporate strategy is executed. *This is what we'll do. This is how it's going to be.* But the lower floors of the prefrontal cortex contain the more primitive machinery of raw appraisal: good or bad, approach or withdraw. This region, spread along the ventral surface of the prefrontal cortex, is called the *orbitofrontal cortex* (because it's just above the orbits of the eyes), and it evaluates what-ever is provided by the senses of vision and hearing, brought in fresh each moment from the back half of the brain, already stamped with limbic meaning, and instantly determines what can be done about it. According to the orbitofrontal cortex (OFC), the world might seem sweet, friendly, safe, hopeful, fulfilling, magnificent, even cautiously optimistic—leading to approach behaviour. Or ominous, threatening, vile, shameful, or uncertain—leading to withdrawal or inhibition, the impulse to stop or to hide. The OFC knows, without conscious judgment, whether the immediate world is attractive or repellent. Then all it has to do is initiate an action *mode*, to advance or retreat, while the executive offices located (dorsally) above it calculate the larger outlook and approve, override, or fine-tune the resulting stream of behaviour.

To understand how the OFC appraises the world—the possible world—and initiates action, we have to look more closely at its henchman: the ventral striatum, nestled just behind it. Here attention is narrowed to specific goals and motivation is whipped up into a froth of forward thrust. Here is where goals are activated and energized, not created, but made manifest, made into acts. Yet in order to link thought with action, to connect the prefrontal cortex with thrust, there has to be a lot of two-way communication between the orbitofrontal cortex and the striatum. These *orbitostriatal* circuits are the crucible of focused action. The thrust that moves us from wanting to getting, from urge to deed. But how do they do it? In order to move us forward, these circuits need fuel, an electrochemical currency, a *neuromodulator* (which is a specialized neurotransmitter that modulates brain activity). And the neuromodulator they need is dopamine.

Dopamine evolved over the last few hundred million years as the fuel for intentional action. It's both high-octane and renewable, so evolution decreed that almost everything to do with action requires dopamine. People with Parkinson's disease lose dopamine, until their capacity to move starts to disintegrate. Obsessive-compulsives are in the throes of too much dopamine, arousing actions they can't turn off. Sex—although it gets launched through another set of chemicals—needs dopamine to energize it. And the forward thrust of action doesn't always come from a "have to"—a desperate need; it can also come from a "want to," slowly simmering in an afternoon's reverie. Where do you draw the line, anyway? Don't ask a lover, an addict, or a two-year-old with his eye on the cookie jar to tell you the difference between need and desire, "have to" and "want to." The thing they desire appropriates their thoughts and magnifies their perceptions, leading to the pursuit, the chase, the anticipation of acquisition, the dopamine rush of thrusting forward in space and time toward, toward, toward . . . whatever it is. They can't stop thinking, focusing, trying, wanting, and eventually, with any luck, *getting* the reward that is etched at the crosshairs of their attention. Which sounds a lot like need. And I could not stop thinking, focusing, trying, wanting, and eventually *getting* Lisa.

With every letter she wrote me, the dopamine pump got activated. And that gave rise to yearning, wanting, and the foretaste of getting. Her pursed lips, the tender symmetry of her features, the pressure of her arms, needing me, wanting *me*, the surrender of her torso to *my* groping arms, the demand of her unwavering gaze . . . these details, coalescing from memory, converge in my orbitofrontal cortex, which, like God's appraisal after six busy days of creation, sees that "It is good." *So good!* Dopamine surging along storm-flooded pipes. Pipes bringing dopamine up to the OFC: oh very, very good. The OFC returning its appraisal back to the striatum, with a fresh ladle of dopamine, buckets from a well. Come on, got to move. Got to write her, call her, imagine her, imagine groping her, got to get to New York. More dopamine sucked up the pipes into the striatum until it is humming with purpose. Molecules of dopamine flooding my synapses, whipping my pulsating axons into further frenzy. Synapse to synapse, this current now carries instructions—steps of action, sequences of steps—around a race track, ending up at the muscle-specific regions of the motor cortex, where they can be articulated into the movements of finger, hand, torso, and tongue, making my body an instrument of intentionality, making things happen in my world. My pen flies over another sheet of stationery. And the rest of Tabor can go to hell; so can my parents, so can Schwartz, so can my probation officer. I just have to see her again.

Dopamine gives me the strength I need to vanquish my depression. And Lisa gives me dopamine. In mid-March the ice in the harbour has melted enough to permit me to float in a small rowboat I've borrowed surreptitiously from the sailing shed. I go out in the waves late in the afternoon, in the slanting sun, shivering with cold and excitement, and tear open today's letter.

"Dear Marc. You're all I care about. My parents are so blind, not to realize how in love we are. They know there's something between us. But they don't see it for what it is. They still think you're kind of a bad boy, you know. If only they could see what I see. I can picture you now. Your soft, warm eyes, your gaze, so intense with love and gentleness and sadness. I don't even know what you see in me . . ."

What do I see? What I see in you, dear Lisa, is nothing but a surge of raw dopamine. That's all I've ever known of you. What I saw in Lisa was the brain state that presented itself as a blissful alternative to grinding depression. Getting and wanting and wanting and getting.

School finally ended in early June. Relief poured through me in a warm torrent. I had convinced my parents and Lisa's parents to let me spend a few weeks in New York before our planned move to California. I visited my probation officer for the last time. And then a brief appearance before the judge. My behaviour was beyond reproach. My sentence was terminated. All the dark clots of dread that had accumulated in the past months now vanished in the haze of early summer.

Day after day, Lisa and I walked and talked. And talked. And talked. As the days lengthened, we roamed New York in search of new adventures, new plays, unknown parks, undiscovered neighbourhoods, novel stretches of harbourfront or skyline. But our conversations grew patchy and strained. Silences lengthened and curled at the edges. What was happening?

Night was necking time, and Lisa finally let me get my hands inside her underpants. My fingertips explored this strange terrain. But the hours on the sofa grew bland and mechanical. My desire to get Lisa's pants all the way off, not just to her knees but all the way onto the floor, had been painfully persistent for the first week. Now it began to subside. Lisa loved me, as she declared with increasing rapture each day. But my feelings for her slackened, steadily, inevitably, leaving a nagging irritation as I lay in the guest bed, trying to make sense of it all. What had happened? How could I stop loving her, now that I finally had her to myself? Surprise turned to guilt. Guilt turned to shame. What kind of person would turn away from this loyal and loving girl, this girl who had seen me through the last months of my imprisonment?

But the answer is simple. The dopamine was gone. Instead of wanting and getting, I was getting, and getting, and getting. Too much getting, not enough wanting. I sat gorged beside an enormous cookie jar, sated and exhausted in my harem of one. My dopamine synapses were drying up and hardening, ceasing to respond to thoughts and images of Lisa. My orbitofrontal circuits, having grown like weeds with images of Lisa as *pure value*, were now overripe, gone to seed, losing the thread of this particular treasure hunt. I would try to go for walks alone, to *be* alone, to rediscover the urge to come back to her. I just wanted to feel that urge, that striatal eruption, once again. But she insisted on being with me all the time. I couldn't get away. Dopamine production turns on when the animal is cued by a stimulus paired with a *likely* reward, by the anticipation of a *possible* reward, or by an increase in the *amount* of reward available. But in the presence of the same old reward time after time, dopamine levels go down! You don't need dopamine once the reward is a certainty, a done deal. So, this particular teenage animal went into what is gracelessly referred to by clinical psychologists as anhedonia—an absence of hedonic feeling. Like mussels dying on a tidal shelf, my orbitostriatal neurons receded and lay still, gaping indifferently at the pale blue sky.

Instead of love, shame now filled my emotion vats. Not unfamiliar, but truly disheartening in this summer of new freedoms. I did not feel proud, or noble, or heroic, or young, attractive, powerful, and victorious. I didn't even feel naughty in that delicious way I had come to relish. Being with Lisa no longer qualified for mythical status—either heroic or villainous. Being with Lisa was *boring*. In the sporadic but relentless mudslide of self-contempt that characterized those years, I decided that, not only wasn't I good in the sense of being a good boy, but I also wasn't good in the sense of being a good person.

PART TWO

LIFE AND DEATH
IN CALIFORNIA

PULLING OUT THE STOPS

A h, Berkeley. You waited for me.

Mom, Dad, Michael, and I drove across the country in two cars, with a moving van somewhere up ahead, the only memorable stop being a night in Lincoln, Nebraska, where we went to see Kubrick's *2001: A Space Odyssey* the week it came out. It seemed like the invariably blonde people of Nebraska weren't grooving to the psychedelic imagery on the screen—imagery that spoke to me like a giant billboard: This way to California! More to come! My parents obviously didn't get it, and Michael was too young. Give him another year. I smoked a relatively weak joint at the far end of the parking lot, very careful now about the inevitable aroma, and I received permission to skip dinner and watch the movie a second time through. I wanted to absorb it completely. In a few more days, we drove across the Bay Bridge into San Francisco and found our apartment on Sacramento Street, near the Children's Hospital. While I languished at Tabor, my dad had switched careers, leaving the leather business for a new preoccupation with the human hide and the organs within it. He had just finished four gruelling years of medical school, much of it at night, and he was about to take up his internship here in San Francisco. But my father's career wasn't high on my list of concerns this week. I wasn't really sure why my parents had decided to move to California, and I didn't much care. I only knew that I was overjoyed to be here. I was taken with San Francisco from the moment I saw its gleaming towers and multicoloured houses, crossing the bridge that first day, the air impossibly bright and clear.

I stayed with my family on Sacramento Street for most of that summer, but I spent more and more time across the bay, in Berkeley. Thanks in large part to Thomas Gelsthorpe. I first called him within a few days of our arrival. I had guarded his address and phone number on a crumpled piece of paper I'd kept in my pocket since leaving Tabor. His voice on the phone sounded eerie — its familiarity a product of myth more than memory — but his words were welcoming. He said to come over the day after tomorrow. He lived six blocks from campus. There was to be a big party or demonstration or something, nobody knew quite what, on July fourth, Independence Day. He said to come by his house and then we'd go together. We'd mingle with the masses of freaks and we would protest the U.S. war in Vietnam. Not only that, but we'd celebrate our own independence — from Tabor Academy, Marion, Massachusetts, hellhole of oppression for the literate and enlightened.

Two days later, my dad dropped me off at the bus station. I found my way to Berkeley with little mishap, but I needed to ask directions a few times to get to Carleton Street. Nobody minded talking to me. Everyone seemed to ooze goodwill. The legendary friendliness of the West Coast was real. And finally, there it was, the dilapidated Victorian he'd described — and there he was on its front porch.

"Welcome, Marse, my boy." This was to be my name for years to come, a product of Tom's compulsion to modify words at every opportunity. "A pleasure to see you, a pleasure of the greatest magnitude," he intoned in a high-pitched, exaggerated humbug of self-parody. We hugged awkwardly. His hair flowed through my fingers.

He wore round glasses that brought out his more effeminate features: round cheeks and a soft chin, augmented by long golden hair, triumphantly exceeding the collar-length limits of Tabor. All crowned by a high forehead that suggested deep and penetrating thoughts. A deep thinker about shallow problems. That's how he presented himself. And yet he was magisterial in my eyes. He'd not only escaped from Tabor, he'd gotten through a year of university. He lived on his own. And he had dope connections! He now gazed at me in his myopic, confused manner, from the litter-strewn foyer of his shared apartment. He seemed perplexed and yet innocently happy. A goofy, brilliant, golden-haired angel. I stared at him with self-conscious delight. Did he mind me showing up at his door? Did he know he represented—in fact embodied—the end of my journey? Me—a frightfully unstable refugee from his own disheartening past?

"Thomas, my good fellow!" I mimicked his genre unconsciously. "You said to come, and here I am. I can't believe I'm here!"

"Oh, believe, believe. You have journeyed far. And you shall be rewarded."

"Rewarded . . . ?"

"Come inside." His voice was weirdly high—like a girl's, I thought. "You never know what the neighbours are thinking." He gazed about nervously. "And before we do anything, I suggest we smoke a joint."

"A joint?"

"Of course we should, Marse, my boy. Of course. We're in California."

A half-hour later, our synapses befuddled by cannabinoids, we walked through the south-campus neighbourhood of slightly rundown Victorians interspersed with stucco monstrosities from the fifties. Most were half-hidden behind clumps of exuberant plants and trees of every variety. Flowers galore, some reaching sunward, bright orange or white and fancifully wild, strutting their stuff. Some resembling red and purple lanterns hanging delicately from their stalks. Redwoods popping up in small thickets in the most improbable spots. Many of the houses were painted in pastels. These had long ago been dissected into student apartments, three or four to a house, and they looked like their inhabitants, searching for new identities but not quite getting them right. Hundred-year-old beige stucco buildings were scattered among them, their small windows full of colour and sound, stereos playing Dylan or the Stones, tie-dyed homemade curtains flapping in the breeze. And there were newer buildings, sixties-modern style, showing white stucco between brown beams, pathetically phony to our stoned yet penetrating eyes. Thomas narrated as we walked. "Plastic, Marse. All plastic. Grown in a vat. Disneyland North. Watch them when you're on acid and you'll see they're not even real." Yet nearly every structure, from dirty grey apartment buildings to seedy mansions with mahogany beams, was adorned, somewhere—on doorframes or gates, in the tiles of the front stoop, in the leaded glass of improbably shaped windows—with art deco emblems, mandalas, or floral curlicues painstakingly coloured by some hippie homeowner.

We reached Telegraph Avenue and turned left toward campus. The air was bright and warm, yet studded here and there with little patches of fog. Some of these patches grew dark as they spread behind the cupolas and trees of Berkeley, reminders of a depression that still might blot out the day's lighthearted mood. Especially to the west, the fog coalesced into a grey horizon at the bottom of each street, so you couldn't see the bay and the distant towers of San Francisco beyond it. But eastward, upward, the Berkeley Hills filled a glorious blue sky, stretching out of sight in both directions.

Telegraph Avenue in those days was a canyon of shops and sidewalk merchants with a tide of moving people and deadlocked cars churning between its walls. The shops included giant record stores and tiny independents selling the latest rock 'n' roll, acid rock, folk rock. And there were clothing shops for every political and social affiliation, including the shawls and scarves that said you were cool, a freak, one of the beautiful people, a hippie, one of *the people*, regular shops that sold blue jeans and their accoutrements, then UC outlets and other super-straight businesses that catered to the 85 percent of Berkeley students who still wore their hair short and had little interest in the war in Vietnam. But among these establishments, like weeds from the sidewalk, appeared a new breed of shops whose front windows and display cases budded with hash pipes, beautifully carved from driftwood or shaped out of ceramics, or blown glass, or copper tubing, under countertops sporting boxes of Zig-Zag papers in every colour and flavour. Racks of Zap comics displaying unspeakable images: spindly naked men with chicken-leg limbs attempting to mount gigantic creatures—half woman, half bird—with enormous breasts and lascivious smiles on their beaks. And boxes of every size and shape: gorgeously painted papier mâché from India, or Afghani silver inlayed with stones and fake jewels, all for storing your psychotropic treasures. There were restaurants swarming with people who carried huge salads overrun with bean sprouts back to long wooden tables and benches, after they'd paid the hairy, good-natured cashiers and exchanged the warmest, the friendliest of intimacies, the wide, sunny smiles of brothers and sisters bound together by the *movement*, whatever the movement was. All of this hinting at a seismic shift of cultures that you couldn't quite put your finger on but could feel in the vibration of the earth itself. We walked past the wafting incense of Shambhala Books, where the posters on the walls and the covers of the books were festooned with mandalas, pulling in your eye, and even your consciousness, until Gelsthorpe said, "Come on, Marse. We can come back here anytime. Let's get to Sproul Plaza, where the action is."

But I could hardly move. Layered in front of the stores were the craftsmen and women with their blankets and tables, and their baskets and felt-covered boards of cheap jewellery, strings of beads of any length, any size, any colour. Beads made of seeds, and stones, shells, ceramic baubles, tiny and graceful or huge and gaudy. It was chaos, with the craftspeople fighting each other in exuberant arguments for turf, and the store owners trying in vain to shoo them away, the crafts-people ignoring them completely, sometimes staring down intently at some nascent objet d'art, hypnotized by their craft or by the joints passing up and down the street. Broad-brimmed leather hats signify-ing different details of coolness according to race. White hippies, hints of rural ethos, good with tools. Or black and blasé, seen it all, been through it all. Kerchiefs meant you were the kind of cool that could not get through another hour without smoking a joint. Motorcycle guys and their big mamas, fat and tattooed, with dark, dirty hands and faces. Scary, but also twinkling. And the spare-changers, the lollers, the leaners, the stoners, the ones who looked half-dead.

Thomas was my guide on this, my first visit to the kingdom, and my face must have registered enough happy innocence to arouse the hungry hordes around me, to let them know that I was fresh off the boat, with full pockets and an open mind. But that wasn't going to last very long, because this jungle of free enterprise didn't stop with bangles and baubles for adorning your wrists or your ear lobes. Everything was for sale on Telegraph Avenue. Not just everything you might want to try on, hang up, or give your girlfriend for her birthday. But everything you might, in the excitement of the moment, in the forge of remaking yourself, want to swallow, like Alice in Wonderland, and wait to see what happened next. This was the layer of Telegraph Avenue that I most wanted to explore . . . had wanted to explore long before I'd said goodbye to Tabor.

"Acid, acid," intoned a skinny man hunched over his knees, leaning against the wall of a storefront, vacant eyes searching mine.

"Orange wedge. Orange wedge."

"Thomas, what's that?"

"It's very strong, Marse. Combines acid and STP. Lasts nearly twenty-four hours, so they tell me. But let's keep walking . . ."

"Blue cheer."

"Purples. I've got purples."

"Reds. Hey man, really good reds."

"Grass, hash, acid. Grass, hash, acid."

"Thomas!" I turned to him repeatedly. "Are these? I mean, is it really . . . ?"

"Certainly, Marse, my boy. The products are very often genuine. Though of course there is no guarantee."

"Do you think . . . I mean, could we maybe?"

Thomas sighed with a little half-smile. "Aren't we impatient today? Well, Marse. If you insist. But I would avoid the fellow with the long ponytail." Indeed, a very skinny man with hollow eyes was lurching toward us.

"No, thank you," Thomas said over his shoulder. And then, a few moments later: "Now there's a trustworthy chap. He's here every day. Give me a minute." And he disappeared into the throng while I gazed around, awestruck and happy.

"What did you get?" I asked when he reappeared.

He opened his palm to reveal a small pill.

"It's purple. Is it acid?"

"I very much hope so," he replied. "In fact it's advertised as Purple Haze." He smiled. "You owe me two dollars." Then he looked at me somberly. "You've never done acid before?"

"No."

"Well, perhaps we should wait till after the demonstration. Things can get a little intense."

And indeed they did. A great crush of people poured into Sproul Plaza, a square of tarmac stretched between the neoclassical pillars of the administration building and the shady steps of the student union. Various people stood on homemade platforms or benches, yelling and cajoling the crowd with loudspeakers. Lots about "The State" and the military-industrial complex. The Vietnam War seemed to come up every few sentences. But I could hardly hear the words above the roar.

"Whose war is it? Is it our war?" A weak chorus of no's from somewhere. "Or is it the war of the rich? The weapon makers! The money makers! The rulers of America want to rule the world! They say they hate communism! I say they hate freedom!"

But hardly anyone was listening. Everyone was having a good time. Knots of people passed joints around. Raucous conversation and laughter rose up everywhere. The hordes blended and flowed in all directions. Nobody was going anywhere. A juggler stood and tossed his balls. People wore clown suits or had their faces painted. And some looked really crazy, at least to me, mumbling nonsense to anyone or no one. It was a carnival. Right here in the middle of the university campus in Berkeley, California. And why not? It was the Fourth of July, and people were celebrating. I was high on cannabinoids and I felt terrific. School was out. Really out.

When the sky reddened, Thomas and I returned to his apartment, and before long his living room was full of people.

"This is Marse, who just escaped from the East Coast. This is Viking. This is Turtle. This is Speedy." What strange names. I felt awkward and I hung back. But wine came out and passed from hand to hand. No glasses necessary. The wine soothed me. A Monopoly board appeared and we began to play, but what I really wanted was what I had wanted all day, what I had wanted for months now. I wanted to try LSD.

"Thomas, do you think we could . . . you know."

"Soon, Marse, soon."

And again a half-hour later. "What about now? Now's a good time, isn't it?"

Finally I wore him down, and he gave me the little purple pill. "What about you?" I asked. "Aren't you going to do it with me?"

"I have guests to attend to, but everyone is cool here. You'll be fine." Down it went, and I sat back next to the Monopoly board, watching houses going up on everyone's properties, feeling light-headed and disoriented, but very excited now. The pill was in me, and all I had to do was wait.

And wait, and wait, and wait. Time went by slowly. Thomas and I exchanged glances every few minutes. He looked bemused, maybe slightly irritated, or was that my imagination? I probably looked like the novice I was. And terribly impatient. When does it start to work?

Nearly forty-five minutes after I swallow the pill, it suddenly begins. The changes are incomprehensible. This is not a drug; it's a switch. Reality breaks apart. The activities going on all around me shift from background to foreground. There is no longer any background. Nothing can be ignored. The hubbub of conversation swells and disintegrates into phrases, each barging in on me, a storm-driven wave. Every face, every gesture takes on enormous power, as if connected to my sensorium through some direct circuit. The room swells and changes in shape and size. It becomes more than a room: an enormous space broken down into subspaces where gripping dramas unfold with each glance, each word spoken or withheld, each facial movement. The skin of those faces decomposes into exotic fabrics made of pores, features, facial hair that seems to grow while I stare, transfixed, horrified. I don't need this much detail. I am overwhelmed by the acceleration itself. Whatever is happening is happening too fast. And a flare of intense physical arousal mushrooms in my belly, keeping time with the sensory meltdown. With each second I feel more suffocated by the enormity of the visual surround, the flood of sound, the welling up of bodily excitement, and the fear of being unable to bear it.

Cannabinoids relax the rules of cortical crowd control, but 300 micrograms of d-lysergic acid diethylamide break them completely. This is a clean sweep. This is the Renaissance after the Dark Ages. Dopamine—the fuel of desire—is only one of four major neuromodulators. Each of the neuromodulators fuels brain operations in its own particular way. But all four of them share two properties. First, they get released and used up all over the brain, not at specific locales. Second, each is produced by one specialized organ, a brain part designed to manufacture that one potent chemical (see Figure 3). Instead of watering the flowers one by one, neuromodulator release is like a sprinkler system. That's why neuromodulators initiate changes that are global, not local. Dopamine fuels attraction, focus, approach, and especially wanting and doing. Norepinephrine fuels perceptual alertness, arousal, excitement, and attention to sensory detail. Acetylcholine energizes all mental operations, consciousness, and thought itself. But the final neuromodulator, serotonin, is more complicated in its action. Serotonin does a lot of different things in a lot of different places, because there are many kinds of serotonin receptors, and they inhabit a great variety of neural nooks, staking out an intricate network.

One of serotonin's most important jobs is to regulate information flow throughout the brain by inhibiting the firing of neurons in many places. And it's the serotonin system that gets dynamited by LSD. Serotonin dampens, it paces, it soothes. It raises the threshold of neurons to the voltage changes induced by glutamate. Remember glutamate? That's the main excitatory neurotransmitter that carries information from synapse to synapse throughout the brain. Serotonin cools this excitation, putting off the next axonal burst, making the receptive neuron less sensitive to the messages it receives from other neurons. Slow down! Take it easy! Don't get carried away by every little molecule of glutamate. Serotonin soothes neurons that might otherwise fire too often, too quickly. If you want to know how it feels to get a serotonin boost, ask a depressive several days into antidepressant therapy. Paxil, Zoloft, Prozac, and all their cousins leave more serotonin in the synapses, hanging around, waiting to help out when the brain becomes too active. Which is most of the time if you feel the world is dark and threatening. Extra serotonin makes the thinking

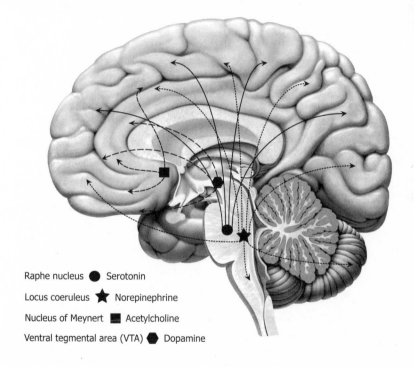

Raphe nucleus ● Serotonin
Locus coeruleus ★ Norepinephrine
Nucleus of Meynert ■ Acetylcholine
Ventral tegmental area (VTA) ⬡ Dopamine

FIGURE 3. A sketch of the four neuromodulator systems of the brain, roughly tracing their source and target locations. Each neuromodulator is fashioned in a single structure, from where it is widely distributed.

process more relaxed—a nice change for depressives, who get a chance to wallow in relative normality.

In a nutshell, serotonin gives your neurons a thick skin, so they can withstand the pace of the bristling, bustling, neural metropolis. And then along comes a tiny army of LSD molecules, marching out of their Trojan Horse—a small purple tablet—and they look just like serotonin molecules. If you were a receptor site, you wouldn't be able to tell the difference. Through this insidious trickery, LSD molecules fool the receptors that normally suck up serotonin. They elbow serotonin out of the way and lodge themselves in these receptors instead. They do this in perceptual regions of the cortex, such as the occipital and temporal lobes, in charge of seeing and hearing, and in more cognitive zones, such as the prefrontal cortex, where conscious judgments take place. They do it in brain-stem nuclei that send their messages throughout the brain and body, felt as arousal and alertness. And once they've taken up their positions, Troy begins to fall. Not through force, as with the devastating blows of alcohol and dextromethorphan, but through passivity. Once encamped in their serotonin receptors, LSD molecules simply remain passive. They don't inhibit, they don't soothe, they don't regulate, or filter, or modulate. They sit back with evil little grins and say, "It's showtime! You just go ahead and fire as much as you like. You're going to pick up a lot of channels you never got before. So have fun. And call me in about eight hours when my shift is over."

What's a brain to do when its blanket of self-regulation is suddenly pulled away, and the impossibly bracing wind of raw sensation bathes its every surface? That brain finds itself attending to everything. Not just what's important, or novel, or meaningful, or expectable, but the tiniest fragments of thought and perception. It is now filled to bursting with images, with information inflow, with detailed inscriptions of a world that was never perceived before.

A world that might never have been perceived with the dark glasses of serotonin firmly in place, protecting its tender pathways. And that brain has three very immediate problems: first, what to do with all that information, all that . . . stuff . . . pouring in through every channel; second, how to fashion some kind of rationality, to make some modicum of sense of the world it now perceives; third, how to control actions, which is the brain's final responsibility. How do you control actions like speaking, listening, going somewhere, even just planning to go somewhere, when so much information is flooding your neurons and there is no particular *reason* to go in any direction?

My brain is reeling without control, without a clue as to how to proceed. I am both preoccupied and overwhelmed with each mushrooming nuance of awareness, propelled without direction by visual displays both frightening and beautiful, by a blizzard of wayward thoughts, and by the unexpected force of my emotions.

I stand up, swaying, and make for the door.

"Where are you going?" asks Thomas, alarmed.

"Don't know. For a walk. Need air." He stares at me for a moment—rising up, swelling—before I bolt out the door and down the stairs, panicked by now, into a foyer that looks twisted and fake, like a cheap prop, but with walls that pour brilliant yellow light into my retinas. And then the light starts to come apart, collapsing into colours like a prism exploding. Red, green, blue, purple, yellow swirls and curlicues, emerging, adhering magically to the complex, mandala-like relief of the textured walls. Without its serotonin filter system, my brain becomes engorged on sensory inflow. It has lost its capacity to integrate. Details explode, each a unique starburst, and I am suffocating, drowning, drowned. Have to keep going. The door to the street opens, and a man and woman come toward me, laughing, joking, now staring at me with enormous beefy faces, huge lips blurting rivers of sound. They are so impossibly *present*. I run past them and finally lunge into the warm night.

I feel enormous relief almost at once. The night embraces me with the gentle movement of air and the sounds of people laughing, dogs barking, all the mundane noises of life. Fragments of conversation coalesce into cartoon dialogues, enacting caricatured scenes, but still recognizable as some variant of Berkeley, of reality. I look out into the dark air, and the scent of flowers penetrates my panic and disarms it. I can feel it receding like a tide. I am all right, and the night is very, very beautiful.

I turn left at the front gate and wander up the street. I propel myself through a pounding undertow of sensory profusion, forcing my feet to walk, one in front of the other, one at a time, past dark gardens pooling with colour and movement. Up ahead, approaching slowly but with melodramatic significance, are the sounds and lights of Telegraph Avenue. First I am frightened by the volume, the crush of people, the speed of cars rushing by, the sheer complexity of the scene around me. But then I see that the sidewalks are half empty, and the traffic is a parade of absurd floats: vehicles designed for parody. Exotic fins from the fifties, clattering bicycles, Volkswagen buses. What Aladdin's lamp is kept buried by serotonin, released by the caress of LSD? Why the sense of reference, the overripe screenplay? Is it all just a Disneyesque default set of connections thrown together by a brain that's lost its momentum, its purchase? Scientists have learned quite a bit about LSD and serotonin receptors, but they haven't come close to figuring out why the world of acid has the character of an absurdist carnival. The groups of colourful people marching along the sidewalk look contrived, like performing seals, laughing and talking, dressed in uniforms of hippie ingenuity, scarves flapping and beads clacking. They are dressed as caricatures of themselves, rendered by the hidden genius of some demented internal artist. Nothing is real. Nothing to get hung about. And so my mood edges back from fear to awe, refreshing itself with each inhalation of the novel world around me.

I am now part of the parade. People in pairs or small groups swell up in front of me and then disappear as suddenly. Their conversations blast into my awareness and then fade to nothing. At first I want to run past them. I am paralyzed with self-consciousness, terrified of meeting their gaze. I stare awkwardly at the sidewalk, but the swirling movement of its texture spooks me more than the people, and my gaze fixes in front of me again. But what if they talk to me? I doubt if I could speak right now. I'm choking on a terribly prominent awareness of the person I try to be, thanks to the runaway firing of cells in circuits of self-monitoring, along the middle walls of my prefrontal cortex, an area that fashions social meaning. And reflections of my body off the dark windows of a furniture store ramify in the sensory regions of my posterior cortex. Who is that absurd-looking mannequin? Meanwhile, in my dorsal-lateral prefrontal cortex, the faculties of judgment obsessively construct *an interpretation* of these suppurating images. I am a human being. One of millions. Part of the river of life. But that's so corny. Those prefrontal cells, dedicated as they are to making sense of things, plunge ahead while the words forming in the back half of my brain are not yet translated. Maybe that's where the absurdity comes in.

I instruct my feet to move me down Telegraph Avenue, toward campus, but anxiety returns like a determined mosquito. Where am I? Where am I going? The sidewalks grow noisier, more crowded, and then suddenly there is nobody there. I look sideways at an overgrown lot, full of garbage—shapes of bottles and broken furniture and unidentifiable things that begin to coil and unwind, separating from their background. Then there's movement right in front of me. I jerk my head around to see, a few metres away, the dark figure of a man standing next to a lamppost, staring. I am transfixed by his face, trying to see it clearly, but I can't. It's too dark, and the shadows below his hat decompose into swaths of colour, now separating from his figure and merging with the grain of the lamppost. His mouth is twitching, his eyes narrowing. Those eyes. They suddenly spark like flint and a voice of unexpected malice freezes my blood.

"What? What did you say?" I stammer, frozen, waiting for more, wanting to run.

"You know," he seems to snarl, yet I don't think he's said anything at all. The words have no echo, no memory. My terror explodes in concentric waves. I need to understand who or what he is.

Then he moves, suddenly like a snake, and I am running so fast, flying down the street, arms wheeling, trying not to fall, convinced that he is right behind me. And I don't stop until my breath is tearing at my lungs and the effort is too great, so I stop running and lurch ahead, walking fast, and I dare to look back but I don't see him, and I keep going, right past Sproul Plaza and into a dark forest of massive trees and buildings—the heart of the campus. I walk, stumble, run a bit, look back, and walk more, but there is no sign of him, and finally I stop completely.

The night rushes in all around me, filling up the void left by my flight with a million twinkling sounds, frogs and running water, crickets, muted voices, and now the low roar of nighttime that weaves them together in a single fabric, nakedly powerful yet peaceful and still.

I look around. I am again struck by the beauty of the night. But I realize with a new stab of panic that I'm completely lost.

I must have looked as lost as I felt, because two people, a man and a woman, came up to me, smiling. Right there in the middle of campus. Their clothes had that flowing, hippie quality I'd come to recognize, replete with a dark red scarf around the woman's neck and a purple vest adorning the man's chest. They were both very attractive.

"Do you need some help?" the man seemed to say.

I tried to answer, but my voice was a blubbering croak, and again came that nauseating wave of self-consciousness. My prefrontal cells normally held on to the phrases of a sentence long enough to formulate an answer. But now they were too busy paying attention to the cascade of colourful outlines radiating from the flowing, cloaked profiles of two such handsome people.

Yet they waited for me patiently, warmly, affectionately. And the elements of speech assembled themselves in my brain and my mouth.

"I'm not sure where to go. And there might be a bad man waiting for me."

"A bad man? Well we can protect you from all the bad men. We're Bonnie and Clyde. Nobody wants to mess with us. What do you say, Bonnie?"

"Sure we can," said the woman. "We're Sal and Al. But you can call us Bonnie and Clyde. You're pretty high, huh?" I had no idea whether they were joking, or in what way they were joking, or who they really were, or whether they were completely real. But I was glad they knew how high I was. Then maybe they could protect me from the chemical bombardment that was all that remained of Berkeley. Maybe they could. And the fact that they *knew* implied a bond between us. A special bond. In fact I was starting to worship them, to love them, more and more each second.

"Where would you like to go?"

"Home!" I blurted.

"Then that's where we'll take you."

"Really? Could you help me find it? There's a man out there who I think wants to hurt me." I had no recollection of having already mentioned him.

"Don't worry," said Al. "You're safe with us."

My guardian angels. Each placed an arm around my shoulders, one on each side, and marched me along the byways of the campus, bathed in the strange light of an orange sky.

I don't think I said much on that journey. Sal and Al chattered with each other, and I remained happily silent between them. In costume and manner they were splendid creatures. They were wild and beautiful, dangerous and kind. They spoke of crazy things: crimes and heists. They really did seem a lot like Bonnie and Clyde. But were they just playing these roles for my sake? I couldn't figure it out.

They must have asked where I'd come from, because we followed my path along Telegraph Avenue in reverse. We got three blocks from campus, to the corner of Haste, when they stopped to chat with friends, and before long we were standing in a widening of the sidewalk among a small throng—eight or ten—passing a joint around. Nobody spoke to me, but I stood close to Sal and Al, just in case anyone tried. Everyone seemed friendly, chatty, connected in some way, part of the night, part of Berkeley, in fact the essence of Berkeley. Bonhomie flowed among them, and I felt its waves lap against my edges, with no need to participate. Just listen, learn, and accept the fact that you have landed in a fairyland. Nobody here wishes you ill. This is *so much better* than anywhere you've ever been before.

And then the police came. Four of them, out of nowhere. *Oh shit.* I stood stock still and scanned people's hands and faces, looking for that joint. But it had miraculously disappeared. So what did they want from us? With their guns and badges and other paraphernalia, they looked like members of some occupying army. They radiated authority . . . but not hostility. In fact they seemed to share some of the spirit of the small crowd, some sense that we were all actors in this classic late-night drama in Berkeley, California, 1968. Of course there were cops. Hippies and cops. Cops and hippies. They were playing their parts so well.

But they wanted something. Anxiety settled in. My breathing echoed in my ears. I wanted to run. Someone said something about a curfew. *What curfew?* The cops were now making their way around the ragged circle, studying the things people had pulled out of their pockets. Wallets, papers. Around the circle, yes, that's what they were doing. Making their way along until . . . they . . . came to . . . *me.* One of them was looking at me, without sympathy or aggression. Just waiting. And I had to do something. But what? My neurons were spectators riveted by this scene of impending doom. The sheer drama lodged itself in my orbitofrontal circuits. There was a lot to take in.

Until the cop in front of me said, "Your ID." Loud and clear. I fished around in my pocket and pulled out my wallet. He looked through it, intently, while silence coalesced.

"Why didn't you speak up when we asked if anyone was under eighteen? Says here you were born in January '51. That makes you seventeen."

"I, uh . . . Um. Not sure . . ."

"Well you'd better be sure. How old are you?"

Awkward pause, and then—it wasn't such a hard question: "Seventeen."

"Where do you live?" That would take a certain amount of thought, and while I considered how to construct my answer, he continued, "And what are you doing here at twelve-thirty, past the curfew?"

This had me stumped. I had no idea what I was doing here. I was the last person to ask what I was doing here. If I tried to explain what I was doing here, it would take hours. Not seeing how to get started, my mouth just hung there. I knew I was in trouble. My vulnerability swelled like a bruise. The cop radiated power. There was no way to stand up to him. No skills, no knowledge, no filters. No filters meant no words, and this was a sticky problem. I was sharply aware that the last time I'd encountered the police had led to a jail cell. I looked toward Sal and Al, and they looked back. I saw that they couldn't save me. Be brave, you have to get through this yourself. But I wasn't quite myself. I was missing a big part of me: the serotonin royal guard, the serotonin phalanx of self-regulation. Lysergic acid was still at the gate, sitting like Yoda with an inscrutable grin that said, "It's your show, baby. You're the composer, the conductor, and a good part of the audience. Do whatever it is you do."

"Did you hear me? What's the problem, boy? You on something?"

The problem, Officer, is that I've lost the capacity to speak. It vanished not long after I took a tablet containing LSD. I'm not sure how you feel about LSD—I suppose that would be the "something" you mentioned—but I'd prefer it if we could avoid the whole subject . . .

We stood facing each other, captured in a freeze-frame. My mouth remained paralyzed. Until time finally started up again.

"Come on," he said wearily to his companions, "we'd better take him in."

Sal and Al gazed after me with concern as I was escorted down the street. But I never saw them after that night. That's one reason why I have never been sure who they were or why they tried to help me. The cops took me to a police station, where I was questioned again, briefly, half-heartedly. They treated me well enough. Here it was, my second bust, and they still didn't seem malevolent to me. Not yet. I got the idea that they knew I was on drugs but weren't going to charge me. They knew, I knew, and we didn't talk about it, just as I'd wished. After the questioning I was escorted to a small room containing nothing but a bench. I was told to sit down and wait: my father was going to come and get me. This did not fully register. I sat and faced the opposite wall. And what a wall it was. For the next hour or so, I watched the wall do things that no wall had ever done—that no wall had a right to do. Its surface became a transparent coat over a multi-layered space, each layer with its own intricate and utterly original pattern, continuously shifting its degree of translucency and changing places with other layers, competing for prominence. And it breathed! The walls breathed! Just like Thomas said they would.

Later, I was able to recall two distinct phases while I sat on that bench. First, I told myself that I was in a jail cell on acid and this was when people would normally freak out. I'd heard about freaking out. That's when people go psychotic and don't come back, sometimes for days or weeks or—who knows?—maybe forever. I really didn't want to freak out. But if I hadn't freaked out yet—and this became more reassuring as the minutes passed—then maybe I wasn't going to freak out. *Ever.* Maybe I would be okay. Even then I had an optimistic view of myself as a career druggie. Then, as anxiety about freaking out faded, anxiety about facing my father grew in its place. Once again, Dad had been summoned to release me from jail. And he wouldn't be happy about it.

I sat. I waited. The walls entertained me. Finally, the door opened and I was taken to meet my dad. In a few more minutes we walked out to his green Alfa Romeo—a recently acquired symbol of his Californian spirit—and got in, without any words. I began to imagine that he didn't know. That the police had told him only that Berkeley had a curfew and I had broken it. It seemed incredible that he didn't notice my state. But he drove without speaking. Maybe he suspected. Or maybe he was totally preoccupied, as was I, by the acceleration to fifth gear on the first span of the nearly empty Bay Bridge. I felt my chest relax, my hands unclench. I wouldn't have to explain the whole LSD thing. I had only one job, I told myself: a task that required all my efforts during the drive home. *Keep all the acid on your side of the car.* Don't let any cross over. Be careful, because it can easily eat through the barrier—the one that runs down the centre, between us, over the gearshift and hand brake. The sense of acid as a liquid presence in the atmosphere of the car was so compelling, so convincing, that I didn't question its authenticity. And I *was* careful, careful not to speak much and to plan any necessary utterances with care. This was getting easier because my acid trip had reached a plateau. My brain found a temporary logic that seemed to work. In fact I was able to gather my thoughts enough to see that things had turned out all right, that this episode was moving toward a happy ending, that I was not in deep shit after all, that acid was really the most amazing thing ever, that I loved Berkeley and everyone in it: even the cops . . . even my dad.

PSYCHEDELICS, SEX, AND VIOLENCE

Every Monday and Thursday night, except when we got our days mixed up, Thomas and I would smoke several bowls of hash and execute a shopping trip to Penny Saver supermarket. We had moved in together in early September, just before classes began. We now shared a rundown one-bedroom apartment on Haste, five blocks west of Telegraph, and the rent was cheap enough to leave money for groceries and drugs out of the $120 a month I got from my folks. I loved our apartment. Two palm trees stood at attention on the front lawn, one on either side of the narrow walk. Palm trees! A few months ago I'd been a prisoner at Tabor Academy. Now I was living in California and attending university. I had been accepted at Berkeley. The gates of the kingdom had been thrown open to this eager-to-please refugee from the East Coast. I was accepted, included, incorporated, in the quasi-hippie subpopulation of Berkeley student life. And I was grateful. I had picked my courses with a mixture of awe and abandon. English, Philosophy, Introductory Psychology, and Astronomy. Why not? I practically skipped to campus in the morning sunshine, high on incredulity. And then, late in the day, after my last class, after lounging about in Sproul Plaza, gazing at pretty women, strolling along Telegraph, buffeted by the hordes, I'd make my way down Haste to Our House. And Thomas would be sitting, waiting, as reliable as an old cat, reading strange comics or Dostoyevsky or Allen Ginsberg. And then, if it was Monday or Thursday, we'd go shopping at Penny Saver.

I developed great affection for our hash pipe. It was sanded and oiled, smooth, burnished wood, with the grain parting gracefully around the little screened bowl. How sweet it felt in my hand, how comforting, but also special, enlightening, bestowing on its host the gift of its magical properties. How could something so benign pack such a wallop? And it became smoother and more comforting, its wood surface absorbing our chemically altered sweat, each time we used it. Most nights, even when we didn't go shopping, we would carve a few fragments off the quarter-ounce of Lebanese Blonde we'd purchased with our pooled resources, carefully, religiously, and feed them into the bowl, into our cheeks, and thence into our cannabinoid receptors. The torrent of babbling neurons would begin, fashioning scenarios of great profundity and hilarity, novel perspectives, and new philosophies on every subject. And then we'd go for a walk somewhere, Thomas with his blond hair in a ponytail and me with my curls gradually growing into a sphere, a halo, a trademark of my new identity.

While smoking dope soon became routine, smoking dope and going to Penny Saver was always an adventure. The trick was not to get paralyzed by the rows of multicoloured cans; not to gape, as if this were a Warhol exhibit; not to collapse into giggles at the sight of so many rolls of toilet paper; not to panic in the checkout line, frozen with self-consciousness, unable to speak. The trick was to balance anxiety and ecstasy while going about our business, unnoticed, behind enemy lines. And who exactly was the enemy? The managers of Penny Saver, the shoppers with their pastel clothing and noisy children, the cashiers, the police, Governor Ronald Reagan? The bad guys, according to Thomas and his friends, were the "straight people." Or, if not bad, then certainly misguided. We, on the other hand, were the freaks, the *people*, brave explorers of the frontiers of the mind. We had read Huxley's *The Doors of Perception*. We took psychedelic drugs. Our very presence at Penny Saver was a rebellion. We boldly strutted down the aisles of corporate excess, spreading the doctrine of individual expression with each burst of giggles.

Ours was a local chapter of the larger split in American society— maybe a central chapter, a trend-setter, as Berkeley was renowned for in those days. You were either a freak or a straight person. You were either with us or against us. You were part of the solution or part of the problem. These slogans bounced off the walls at social gatherings, which often turned out to be political gatherings. It was the era of *Easy Rider*, Spiro Agnew, and "America: love it or leave it." Kids got beaten up now and then for their long hair or strident views. Neither Thomas nor I were serious enough for any real political action, or even political thought, but we took up the banner of freakdom with enthusiasm. With reverence. We liked our drugs, and we liked to dress up. We were hippies in style, if not in substance. We were little boys approaching adulthood with no idea where it lay.

Partying with Thomas became the centre of my life. Our living room, its threadbare Persian carpet scattered with stray Monopoly tokens and chess men, became the place where I felt safest. I was happy and relieved that Thomas actually liked me enough to want to live together, and this made it easier to ignore his odd characteristics—in particular, his flowery falsetto speech and his determination to avoid any sort of serious, straightforward conversation. Only occasionally did he drive me crazy. But most of the time I followed his lead and imitated his carefree air, grateful to be included in this little world of his. It seemed good and right that I lived on Haste Street with Thomas and went to school at Berkeley. These were the trappings of a life, a real grown-up life. A life independent of my parents (except for the cheques that came each month). I didn't need them to take care of me. This was the credo I'd established at Tabor, and now it served me particularly well. In fact my parents seemed less accessible—a mere hour away—than they had during Tabor days. Perhaps due in part to their own problems.

Michael lived with my folks in a rented house in San Francisco—a dark, vine-covered house hunched over a low brick wall surrounding an enormous park. I came to spend the night once a week, feeling more and more like a tourist visiting my old life. Michael and I would wander around the neighbourhood, hypnotized by the magic of the San Francisco night, whose vistas were wet and mystical from just about anywhere. And then, one evening in late October, Mom and Dad announced that we were going to have a little family meeting. We were puzzled. This was not their usual style. They sat us down on the sofa and spoke in parallel. "Your father and I are thinking about getting divorced." Just like that. No shouting or fighting, no emotion that would have grounded the thing in reality. "We're discussing it at this point. It isn't for sure." The pantomime of teamwork took on creepy overtones. When I went home to Thomas that night, it began to dawn on me that no other home might exist.

I took psychedelic drugs two or three times a week, often in the beautiful Berkeley Hills, or wandering around campus, watching the vapour trails of flying geese. Or in Golden Gate Park, or in the night-lit mounds of the Presidio golf course, undulating like pink glaciers, or just wandering the streets, assessing the hieroglyphs of the sidewalk to reassure myself that I was good and high. I smoked pot or hash nearly every day. But everyone smoked dope. It was my fascination with psychedelics—acid, STP, mescaline, and psilocybin—that seemed a little weird, even to my peers. Thomas and his friends would "trip" every few weeks or every few months, but for me the very thought of an acid trip—right here, right now—caused a lurching shift in my dopamine circuitry, where a zoom lens latched onto the goal of getting high. Once I started to think about it, I just couldn't shake the thought. As with Lisa, the excited craving of something half-attainable was the most potent elixir. It's true: my attitude toward LSD, a drug, was not much different from my zeal to connect with Lisa, a girl, thanks to a flood of dopamine in my ventral striatum—wanting and wanting and wanting and finally getting, magnified by the *uncertainty* of the goal. And what could be more uncertain than swallowing an unauthorized, mind-altering drug of unknown dosage in an unfamiliar place in an unfamiliar world? Most of the time, especially if I was feeling insecure, or depressed, or alienated, the attraction would build and build and finally climax into *Oh fuck it*, ending in another ill-planned psychedelic journey, just days after the last one. The objective was to get out there— way, way out there. To go somewhere tremendously new and different, somewhere exciting and a bit scary. I wasn't sure why. Probably to escape the claustrophobic monotony of being me.

And I got into various kinds of trouble.

I had spent one acid trip late in the summer sitting and sunning myself at a folk concert on the steps of Sproul Hall. When evening came, I stuck out my thumb and got a ride from a man who soon made it clear that he wanted to have sex with me. There wasn't enough serotonin in my synapses to ride out the ensuing waves of panic. I began to go rigid with fear in the front seat of his car, inches away from his hideous body. Was I being kidnapped? How would I escape? I finally got it together enough to say, "This is my stop. I want to get out *here*." *Here* happened to be a highway ramp. But he pulled over. And within minutes I was picked up by the California Highway Patrol for walking along the freeway. It was late; I was still seventeen.

97

Suspected runaway. A whole night in jail this time, waiting for the colours to die down so I could sleep.

Then, in September, two hours into a moderately strong acid trip, I stopped to smoke a joint in a glade on the Berkeley campus, a few paces above the banks of Strawberry Creek. Out of sight of the world, I thought. But a husky young man came up to me and asked if I was smoking something. I would have shared my joint with him, but I had just finished it. I said "No," to be on the safe side, and he told me I was under arrest. I bolted, instantly, without thought. I ran headlong through the glade, trying to outpace him, terrified because I still had one joint sitting in my pocket. He eventually caught up and grabbed me by the neck, moments after I got the pocket open and shoved the joint into my mouth. It wasn't easy to swallow something as big as a joint while running, or while in the grip of a half nelson, applied by a sweating and angry plainclothes cop. But I finally got it down my throat while being goose-stepped to a nearby squad car. I sat for a long while in the back seat, my hands cuffed painfully behind me, in a now-familiar state of near-panic, watching an amazing scene unfold. One after another, four or five police cars pulled up and disgorged a small army of cops who milled about with flashlights, looking for the joint I knew wasn't there. I had swallowed it. I began to smile with private satisfaction. But eventually I was busted for "resisting arrest" and possession of paraphernalia—that being a rolled-up matchbook cover, burnt at one end. More hours spent staring at a wall. And then I was released on my own recognizance, which wasn't very reliable that night.

Then came a nasty misadventure that surely shaped my sense of the world for years to come. And like the others, it started in happenstance and ended in some sort of twisted irony. I had worked that summer for about seven weeks—that's all I'd lasted—in the men's department of a large department store in downtown San Francisco.

And there I was surrounded by a covey of gay men. Gay men were just starting to call themselves "gay" in the late sixties, but in San Francisco the social bravado of the gay subculture was already legendary and growing by the month. I liked these guys. They were intelligent and dry and humorous. They seemed to have a sense of the absurd that filled some of the gaps I found in the adult world around me. I was eager to learn and, as always, to be accepted. I was young, fresh-faced, completely naive, and possibly of interest to some of these men, in a way I'd never before imagined. I didn't consider going out with any of them. In fact nobody asked me. I didn't think I was gay. But there was a flattering sense of being admired that soothed my ongoing identity crisis.

In early November I dropped acid on an otherwise dreary evening. Not too much. The right amount to get high and still have my wits about me. The whorls on the Persian carpet began to shift subtly, suggestively. Thomas was out somewhere. I was restless, excited, lonely, lost in my living room. I decided to go visit Stuart, one of the guys from the store who seemed to like me. I had his address in San Francisco. I called, I walked over to Ashby, and I started hitchhiking. I was in the city twenty-five minutes later, and I worked my way over to North Beach by nine or so.

Stuart came to the door and told me he had guests but I was welcome to come in. He seemed a little distant. I followed him into the living room and sat down, and I soon felt the conversation around me lose its momentum. The guys who were there—all guys—seemed subdued. One asked where I was from. Another asked how I liked Berkeley. Stale questions to ease the silence. I told no one except Stuart that I had taken acid a few hours ago. I could handle it. Nobody could tell. But why were they being like that? I felt a little desperate to be included, treated like one of them.

"But you're *not* one of us," Stuart told me sharply in the kitchen.

"You mean I'm not gay? Or . . ."

"I mean you don't know what you are." I didn't understand what he was saying.

"And you make people uncomfortable." Dread began to sprout in my stomach. "You flirt with us. You flirt with me. And nobody knows whether you mean it or not."

"I . . . flirt?" Wanting very badly to be forgiven, but I had no idea what for. "Stuart, I don't flirt."

"Yes, you do. It's subtle. But it's there. Don't you think we've got radar for young guys like you, aching to be loved? You gaze at me, waiting, all gooey-eyed. And at others too. What are you waiting for, Marc? To be fucked? Are you offering? Nobody's going to come near you until you start to give off clear signals."

I left moments later, leaden with shame. All summer long I'd been flirting with these guys and not knowing it? Did I want to be seduced? Unconsciously? The acid was still strong enough to blur my thinking. What emerged was an image of myself as crassly self-centred, parading myself before . . . *them*. My crime seemed to be my presumption. And wasn't it pathetic to imagine that I was this tolerant, liberal-minded guy who really accepted homosexuals? Yet held myself aloof? A familiar voice disparaged me for being a fake. The sights and sounds of the strip clubs proliferated as I approached Broadway, but my preoccupation with my own ugliness blocked them out.

I hitchhiked to Berkeley and was let off at the foot of University Avenue. And then I started walking, eastward toward campus. It was a long way from the freeway to the little civilization of Victorians that had grown up around campus. I stuck my thumb out every block or so, without much hope. And then a van slowed down just ahead of me. Its door opened up, but I didn't see anyone at first. Were they stopping for me?

It wasn't until I was a few steps from the van's front door that I saw a face peering out at me. It was a black face. My first instinct was to back away. The guy was big and not particularly gentle looking. His little eyes shone balefully out of dark features. Next to him was a thin black man, not as menacing in appearance, but not radiating much kindness, either. Black guys didn't pick up white hitchhikers. They just didn't. A tangible waft of threat came from the open door, and I hung back.

"You want a ride, man?"

"Well, no, not really."

"Then how come you had your thumb out? We saw you. That's why we slowed down. We just bein' friendly. Like. Take it or leave it."

Taunting homosexuals was bad enough. Now I was being a racist as well.

The amygdala sits like a crab in each hemisphere, right and left, hovering over a stream of incoming sensory information from the back of the cortex. It marks this information as emotionally significant—for example, anxious-making—if it matches up with previous events of an emotional nature. The amygdala is the organ of emotional memory, emotional associations. It triggers the whole cascade of neurochemical events that comprise an emotional reaction. It tells you that the movement in the grass could be a snake. But at a level far below thought or even conscious perception. It's an automatic alert flag, to make you sit up and pay attention. And it performs this function by way of its connections downstream: to the brain stem, where norepinephrine gets released like a chemical floodlight, tuning the brain to every perceptual nuance, and to the hypothalamus, which sits on top of the brain stem, coordinating its many functions (see Figure 1). The hypothalamus does a lot of things, but one of them is to activate the sympathetic (fight-or-flight) nervous system—a key component of most emotions. The hypothalamus–brain stem assembly also houses ancient behavioural programs, like freezing or fleeing, fighting or fornicating, preparing the body for whatever dangers or opportunities are at hand. All this ramping up of the lower brain starts with the amygdala's wake-up call. But soon after the amygdala comes on line, so does the orbitofrontal cortex (OFC), where the present threat gets evaluated *in context*, in *relation* to the bigger picture. The OFC, in turn, sends messages back to the amygdala, calming or calibrating its primitive concerns. This is how human intelligence begins to moderate the animal urgency of emotion.

Brain-scan studies over the past decade have pinpointed how white Americans typically react to black faces. In one such study, black faces on screen for just a thirtieth of a second triggered a spike in the amygdala. But when those faces were shown for half a second (a long time for perception), amygdala activation went back to normal and prefrontal activation went up instead. So these people must have felt some emotional reaction they (semiconsciously) deemed racist. Most likely, fear. It's not morally, socially, or politically appropriate to cringe when you see a black face. So they tried to control it—and, judging by the drop in amygdala activity, they succeeded. That's probably just what occurred to me that night in Berkeley. Although I was tired, distraught, and stoned, although I was young and confused, a Canadian roaming around an American city, a preppy trying on the trappings of West Coast hippiedom, I still got the message: Don't be a racist. These guys are just giving you a lift. Be decent, tolerant, enlightened, and accept the ride the nice man is offering you. My prefrontal cortex forged these calculations despite a steady hiss of alarm from my amygdala. I nodded emphatically, thanked the man, and got in.

Which just goes to show that brains don't always know what's good for them.

The big guy on the passenger side pulled his seat forward so I could slip into the back. The back was an empty compartment with a few nameless odds and ends strewn about. The entire interior was covered by a featureless coat of dirty white paint. The floor consisted of corrugated ribs of metal. There was nothing comfortable anywhere. I sat on a wooden crate as the van lurched forward.

"Where you wanna go?"

"Oh, just up University to campus. I can get off there."

"No need. We'll take you where you wanna go." I didn't say anything. I didn't understand the offer. People don't offer to take you anywhere you want to go. They offer to drop you off along their route.

We drove on in silence. But my OFC was completely at a loss to interpret what was happening. My amygdala ramped up its firing rate. There was nothing good here.

"So where did you say you lived?"

"No, it's okay, really. Just . . ."

"Listen, man. I'm askin' you where you live so we can take you there. Now where we gonna take you?"

"No, really. It's okay."

He turned around in his seat and fixed me with a bad-guy stare: "I said, where we gonna take you?"

"I . . ."

"I ain't fuckin' around, boy. Where your house?"

"Well, it's down Grove, but you don't have to . . ."

"Sure we have to," he turned back around. His voice was now soft and oily. I felt sweat on the skin of my hands, my arms, everywhere. My sympathetic nervous system was turned on all the way, my mouth was dry, and my muscles clenched. They could not avoid it. Because my amygdala, hypothalamus, and brain stem were locked together in a dance of raw fear. My amygdala activated circuits in my hypothalamus and brain stem, and they in turn elicited chemical geysers that cranked up all brain and bodily systems tuned to escape. These were reactions I could not control.

The van took a right on Grove and then, "What's the address, man?"

"Look, it's okay . . ." I stammered.

The van veered over to the side of the road and slammed to a halt. "It's not okay," said the driver, "because we's offering you a ride and you won't say where you live." The man on the passenger side had turned around again, all the way this time. He looked very angry. And then his face softened.

"Hey, you got any dope, man?"

"No, I don't." My heart rate continued to climb. The sympathetic nervous system includes the adrenal glands, which now released torrents of adrenalin. Adrenalin got sucked up by receptors all over my body: stomach, sweat glands, heart, lungs. My heart was rocketing, preparing for the worst. Its beating was thunderous.

"What about at home? You got some dope at home?" No way was I going to take them to my house.

"I want to get out here," I croaked. And then he started to rise up out of his seat. I bolted for the rear door of the van. I threw my shoulder against it but it didn't budge. I tried desperately to rotate the handle, but it was locked in place. And then he was on me. He punched me in the face and I slammed against the side of the van. He was incredibly strong. The other man's head appeared, hovering behind the shoulder of my attacker. Both were hunched below the low ceiling. I was on my knees against the metal wall. I tried to get up and push the big one away from me, but this enraged him further. "Racist asshole," he sneered, and then hit me again, and then again, each blow powerful enough to throw my body back into the wall. Yet I could feel little pain. Some part of me was turning off, going dormant. The attack slowed down, stopped, reversed, transformed into fake friendliness—a drawn-out, one-sided conversation in which I pretended to see the error of my ways. Shame soaked through my numbness. And then a stream of insults would emit from his twisted lips, marking the start of another round. I was a racist asshole because I didn't want black people in my house. He said this almost tenderly, holding me close to him. And then his fist would slam into my stomach, his knee into my groin. I was a motherfucking racist. And I was being punished.

The last thing he said before tossing me from the van was that they'd be back. They'd be in Berkeley again in a few months. And they would come looking for me. And if they found me, I would have to show them I'd learned some respect, some hospitality. Or else they'd give me another lesson.

Several weeks later, my face and body had nearly returned to normal. But a dull nausea persisted. It felt more spiritual than physical or mental, as if my soul were ill, and it would not go away. All I wanted was to purge the memory of that night. I had never been hated like that. I had never been hurt like that. And I had capitulated, hadn't I? I didn't fight back, the way a real man would, a guy like Bob Moore or even Miles. Shame and fear merged viciously. They said they'd be back. I couldn't stop thinking about it. I told Thomas, I told my mom, but nobody had any advice to give me. Nobody seemed to get it. I shouldn't have been hitchhiking so late. Whatever. Over and over, shame, fear, and anger coalesced into familiar, terrible thoughts. And my internal gaze, the baleful eye that watched me from somewhere inside, continued its recognizance. Somehow, according to its verdict, I was the one at fault. It was I who deserved rebuke. Stuart told me I was a flirt, a tease, and then my racist thoughts — no, my stupid attempt to *deny* my racism — got me beat up. This mirror of self-blame was familiar from Tabor days. As though some central mechanism of selfhood were simply broken, I could not pinpoint any fault except my own.

MARC LEWIS

———

Berkeley days were now dulled by the rainy season. Outside my window, the skies were grey and featureless. Tom had found a girlfriend. I think her name was Dianne. She was tall, with large cheekbones, a flabby neck, and dark blonde hair. With her glasses and her oddly stiff mannerisms, she could have been his long-lost sister. But he seemed content. We shared the bedroom in our small apartment, with two single beds side by side, and we shared the living room, a 3-D mandala whose centre was the Persian carpet around which we sat to play Monopoly or hallucinate on the purple whorls. This was not an inviting arrangement for Dianne, whose wall-shaking passion was loud enough to serve as a traffic signal for me, the roommate in the wings. I'd wait in the living room, or in the tiny, bright kitchen, until the last tremors had subsided, and then sneak into my bed, whose edge was less than a metre from theirs. I could reach out and put my arm around Dianne almost as easily as Thomas could. And for this reason, I'm sure, it was Dianne's house that became their preferred hangout. Our apartment was mostly empty.

It was in this world, defined by an absent roommate and a residue of undigested violence, that I tried to groom my identity as a young adult: confident, cheerful, sensitive, and yet mature, a comfortable person you'd enjoy being around. But it didn't work. For one thing, I was lonely. I just couldn't seem to find real friends. For another, I was restless. I didn't know what to do with myself. Sit at home and do . . . what? Go out and do . . . what? Go to San Francisco and hang out with my parents? Michael had his own friends now. I didn't feel right tagging along with them. No, it was just me most nights. And you couldn't go on an acid trip any sooner than three days after the last one. Some poorly understood tolerance made the drug nearly ineffective for seventy-two hours before you were good to go again. What I really needed, I told myself again and again, was a girlfriend.

But what did I have to offer a woman? What kind of brain—never mind body—could I put on the table? My serotonin receptors, invaded every three days by LSD molecules, were half-shattered, hungover, and confused. Incoming information was raw and jagged. And the information embodied by my own thoughts, my roiling, untidy, and often downright vicious internal dialogue, filled the empty spaces with spiky vegetation. A no-trespassing zone for any potential girlfriend. My amygdala was on high alert. They said they'd be back. They said they'd look for me. The world was pregnant with danger. The police were not my friends, my gay friends were not my friends, and my parents were about to get a divorce. My apartment, my haven, was a cage suspended above the snapping jaws of unseen predators. There was no safe place, inside or out, and tides of norepinephrine fixed my attention on anything out of the ordinary, novel, potentially dangerous, like the echoey emptiness of the nighttime streets, the rock music booming from the apartment next door—even the thought of sex, so far a complete mystery. I needed soothing. I needed to feel connected. I needed to feel like a man.

My best bet was to find someone who wasn't in great shape either. And so I chose Susan—and she chose me—out of convenience. Susan had small, shy eyes and thin, frail lips. She was as tall as me, a little gawky, a bit like a giraffe, and the skin on her upper arms was chafed with some kind of rash. She spoke softly, as if she were embarrassed by whatever she had to say. In fact, Susan could radiate insecurity. And yet she did not back away from life. She was no beauty queen, but her long dark hair swished with its own mystique, and she faced the world with some kind of big-girl honesty that I found courageous. Susan had a lover, a sporadic Zorro who came riding in on his Triumph Bonneville 650 and spent the night with her—only once or twice a month, as far as I could tell. His name was Peter and he wore leathers: a tan leather jacket, fringed at the edges. With his jingles and jangles and his good looks and smooth manner, Peter was a good catch. But he seemed to have many other calls to make, and Susan wanted someone a little more consistent. I wanted someone who was nice enough, pretty enough, smart enough, just to like. I wanted someone I enjoyed talking with, someone I could feel proud walking down the street with, and Susan would do just fine.

Susan came as part of a package: a box set of lovely Berkeley ladies, flowing, flowering, mysterious, yet capable—an idealized dream of womanhood to my impoverished senses. The four women lived together in a grand old house on Piedmont Street, in the stately but slightly shabby region of Berkeley at the foot of the hills just north of Oakland. The house was painted a soft, minty green, it was surrounded by delicate shrubs and fountains of flowers, and it was often permeated by the smells of vegetarian cooking. Onion pie and stewed eggplant came to life in that house. And so did my libido. I was falling in love with this house and its bevy of beauties, its promise of tantalizing new worlds, and with Susan herself. I came back again and again, like a stray cat, looking for comfort, food, love, and sex.

Sex was of paramount importance. When I met Susan and her housemates, I was still a virgin and not very proud of it. I seemed to be the only virgin in the entire Berkeley area. I was essentially a freak. But it was curable. *I* was curable. And one night after dinner I finally got up the nerve to try.

"Do you suppose you'd like to . . . you and I . . . would you want to . . . ?"

"Want to . . . ?" She smiled softly, with a touch of mockery.

"You know. Sleep together?" There, it's done. But for a moment she looked sad, and that horrified me. Was I asking too much? Too soon? Was I acting like every other guy who just wanted to screw? In such new terrain, my critical sensors were on high alert for misbehaviour. But Susan must have seen what a hopeless novice I was, because she put her hand on my leg and stroked me softly, making it easy. She smiled a smile both wise and fragile.

"Come on," she said and stood up. I stood up and followed. She led me into the house, toward the bedrooms, into her room. Lightly furnished girl room: knick-knacks here and there, coloured prints, tacked-up fabrics, incense holders, scrawny plants. The bed was covered by a tie-dyed cover. She reached down with one hand and moved it aside while we kissed. She knew what she was doing. Soon we were necking seriously on the edge of the bed and then, without hurry or hesitation, she pulled me down and guided me onto her. She reached for the light. Suddenly, there was nothing but shadow, strange sounds, strange smells. I felt afraid. Where is she?! Then her arms were around me and I fell into her, as if her flesh were a nest, a home.

I became very fond of Susan as the weeks went by. The qualities of her Susan-ness, which had seemed so arbitrary at first, now seemed precious and compelling. The cross-hairs of my attention, spiked with dopamine-powered lust, now lingered on the red trellis woven into her white Indian blouse. Her bra-less breasts half-hidden in their nook. Her feet, always bare, tawny and delicate. The flair of her bell-bottoms seemed the epitome of female style. Her candles and incense holders struck me as the humble paraphernalia of her quest for the profound. I admired her. I became familiar with the dip of her chin when she revealed some secret; her awkward shyness was endearing to me. Her long limbs were now enticing more than gawky. Her oval glasses somehow highlighted her sexiness. She seemed demure rather than nervous, feminine rather than withdrawn. And by candle-light, everything about her, from the intimacy of her moist palms to the dark tangle of her braids, breathed softly of sex.

Susan and I were pals, mates, lovers, buddies, fellow travellers. I didn't worship her, but I cared about her. And she didn't worship me, but she was fond of me. She made me as happy as I was capable of being. That she didn't worship me was particularly clear on those occasional nights when I'd find Peter's Triumph parked at her front porch. On those nights I would just stand there for a few minutes, befuddled, angry, and sad, waiting, wanting, craving her and not having anywhere else to go or anyone else to see. But that was Berkeley in those days, and the rules of free love meant I was also free to have sex with others.

I did my best. I had a brief fling with the red-haired woman downstairs, after her boyfriend stuck his head in the oven, thereby terminating their relationship. And there was Sheila the nurse. She and her roommate liked to frolic with Tom and me on the Frisbee field by day and the bedroom by night. I didn't form close bonds with these women. It was Susan I came back to again and again. Susan became an imprint on my nervous system. We hitchhiked together to Big Sur and camped on the beach. We dropped acid in the Berkeley Hills. We escaped hand in hand from a gang of drunken Marines. We demonstrated together against the war in Vietnam, against the university's takeover of People's Park. We drank wine on top of Mount Tamalpais, overlooking a hundred kilometres of Pacific coastline. She was often with me. In daytime and at night. I dreamed about her then and for years afterward. Not exactly her, but the figure of a fragile young woman, whose essence had infiltrated my images of all women, and whose unstinting affection kept me feeling safe in a perilous world.

It was with Susan that I got my first shuddery taste of the destructive power of psychedelic drugs. I had got a new batch of acid and I wanted to try it with her. She was not nearly as fond of psychedelics as I was. They made her nervous, and I only understood why after it was too late. But this night she consented. We swallowed our pills at her place. We started to feel the first waves of energy while sitting in her living room. The aromas of flowers and baking bread took on extraordinary prominence. I was exuberant. She was uncomfortable.

"Aren't you having fun?" I asked.

"It's so . . . sudden," she said, "and I don't want to be around other people. I don't want to . . . to have to . . . talk to anyone."

"Then let's go to my house! Thomas won't be there. For sure. I know he's out of town. It'll be just us." But getting there was problematic. It was a long walk, and I didn't think Susan would make it. My bicycle was right outside. I'd ride her double. It was downhill nearly all the way.

I thought this was a brilliant plan, but Susan was far from convinced. I had to pull her along, cajole her, convince her, and I told myself I had to do it fast. The acid had just begun to come on, and it would get stronger, moment by moment, for about another hour. Better to be there, safe and sound, than to be caught on the street when it hit its stride.

So I just about dragged her out to the front, got her onto the seat, and got the bike going, wobbling like crazy until I swung my leg over the cross-bar and stabilized its course. I thought we were doing okay. I was able to handle the bike, despite her weight, as we coasted down Piedmont toward Parker. I took a quick left and felt her hands dig into me, too hard, hurting me. I asked her to take it easy. This wasn't helping. But she did not reply and her grip didn't change. It was too difficult to turn around and look at her while speeding down the street. I assumed she was just nervous. I kept up a banter of soothing words, as much as I could in the grip of the drug. This stuff was coming on fast. And it was strong! The outlines of cars separated into multicoloured extensions, phantom car bodies behind and in front of them. Steering the bike was a game of pinball, a careening, barely controlled spillway through corridors of shimmering metal. I was absolutely absorbed in the challenge and the thrill of it. I had no idea how badly frightened Susan was behind me.

We got there in one piece, but Susan's face froze me. She looked hypnotized, tranced out, not at all what I expected. I thought she'd be stuttering with anxiety, maybe pissed off, maybe just desperate to get inside. But her face showed no expression at all. It was as if she weren't there inside her body and all I was seeing was a shell.

I walked her up the stairs, quickly but gently. She'd be okay. I put on the stereo: something soft and soothing. That always worked for me when I was too high. But it didn't work. Susan sat stiffly where I had placed her, in our one proper easy chair. She looked straight ahead. She did not respond to anything I said. She would not talk. I went into the kitchen to get her some wine. I thought that would calm her down. But when I came back she was on the floor, in a corner of the room. She was half-collapsed against a bookcase, passing her fingertips over the spines of the books as though she had no idea what they were.

"Susan, what's going on?" I asked, half-panicked myself.

Her reply was a series of mewling sounds. She did not look at me. She began to lick her fingers, repetitively, compulsively.

"Are you an animal?" I tried to joke.

But to this she finally replied. "I'm a kitten. Do you want to play?"

She glanced up briefly, not seeming to recognize me. Her eyes were glassy and set. She continued to lick herself, and then she crawled a little farther along the wall, curled up, and started whining. It took me a few tries before I fully realized that this was not pretend. She was in another world. She was completely lost to me, and seemingly to herself as well. She had no idea who she was or where she was, let alone who I was. I was very frightened, and though I was high on acid, the urgency of the situation made it necessary to think.

There were a few folk remedies thought to counteract an over-dose of LSD. Orange juice was one. But as far as I knew, the most reliable antidote was barbiturate, the purest form of downer. At the root of the terror that attacks the psyche is the excessive glut of infor-mation, both sensory and mental, normally blocked by serotonin. Susan, I thought, was drowning in some kind of psychological excess, and barbiturate would reduce it to bearable levels. She needed to be sedated.

But where would I get any sort of barbiturate, especially at this hour?

I phoned Michael in San Francisco. I told him this was a real emergency, and he had to help me. He was to go to my parents' medicine chest and get some Seconal. I knew my dad took Seconal for insomnia. Michael was to grab a few and get to Berkeley right away.

"Take a taxi to the bus station. Take the bus to Ashby and Grove. Then try to flag another cab. Don't worry, I'll pay you back."

"But what if I don't . . ."

"I don't care! Get the money from Mom. Tell her . . . tell her I need . . . I need . . . bring a couple of books. Tell her I need some books. Tell her I've got an exam tomorrow."

Michael must have heard my panic, because he did everything I asked. He arrived at the door, flushed and scared, less than an hour later. Susan's condition hadn't changed. She was still on the floor, still making animal sounds. Still completely unaware of where we were and what was going on. I believed she could remain that way forever.

"These are pills especially for cats!" I improvised. "We're all going to take some, and then we're going to play together." At first she pushed the small red capsules away, but Michael and I persisted, we pretended to take a couple, and we wore her down. We finally got two down her throat. Then we waited.

The inhibitory neurotransmitter GABA is greatly enhanced by barbiturates. That's how barbiturates work. In less than twenty minutes, Susan's brain started to decelerate. Within half an hour, the data stream became manageable. Before long, the background noise got low enough to permit "reality" to peek through. This was not a pleasant transition. She began to look tense, then frightened, then terrified. It was as if she were moving through time in reverse, feeling the feelings that must have overwhelmed her hours before. Yet it wasn't just the terror on her face that I found so disturbing. It was the look of awestruck realization: the horror of coming back to reality and *realizing* just how completely she'd left it. Like waking from a coma. It was as though she had lived in a dream for so long that it had become her entire world. She forgot there'd ever been any other world. But now the dream was crumbling, that cozy world was splayed open, and beneath its surface lay a festering, larval nest of pure insanity. That was the home she'd built when her brain got too stretched. She had gone psychotic, and now she was crawling back to reality. Slowly, painfully. And it wasn't just a harsher world than the one she'd invented; it was a world whose latest news concerned her own fragmentation, her complete and utter frailty as a human being. She had let herself fall apart.

7
A PSYCHEDELIC FINALE:
COPS AND ANGELS

In the spring of 1969, Governor Ronald Reagan decided to invade "People's Park" and convert it to what it had long been destined for: a parking lot. Rumours of this move had been circulating for months, and they had a profound effect on the residents of south Berkeley. Here had been a vacant lot that took up most of a square city block, one street east of Telegraph Avenue, four blocks south of campus. A few hippies slept there in tents or on mattresses, people grew vegetables in spindly troughs, and probably a fair bit of drug dealing went on. But the land belonged to the University of California. And when people discovered that the university planned to reclaim it, some weird sort of alchemy transformed it into an idealized wonderland of hippiedom.

"Have you heard? The pigs are going to shut down People's Park."

"What? They can't do that!"

"Oh yes they can. They're going to pave it over. It's going to be a parking lot, just like everything else."

"But what about the gardens? What about the people who sleep there?"

"They couldn't care less. It's their land, and they're going to claim it."

Conversations like this were heard often, and with them came action. Wooden playground structures were erected. A well-groomed vegetable garden appeared from nowhere. Flowers blossomed. Bushes and shrubs appeared. An oasis grew out of the desert. And the place magically cleaned itself up. Mattresses and other artifacts of permanence were removed. A modest community of tents was allowed to remain. Bands played while people worked to fix up the park. It was lovely! I didn't know who was calling the shots, but People's Park grew, through this infusion of political and communal idealism, into a beautiful place: a glorious manifestation of squatters' rights.

The university was not impressed. Talks and meetings were held. A compromise solution was sought. But it turned out that Reagan had been itching for a showdown with the freaks, weirdos, and communists of Berkeley, ever since the jewel of the University of California became a haven for political protest and free speech. So, while negotiations were stalled between the university and the community, the State of California made its move. The California Highway Patrol and Berkeley police joined forces and came by night to vapourize the park. They destroyed most of the gardens and removed the structures. They either chased the transients away or locked them up—no one knew. The next morning, we woke to find the park surrounded by a high chain-link fence. Looking through it, one could see what looked like a wasteland, just as it had been before. The park was gone. And the streets were full of police.

That's when the riots started. A rally in Sproul Plaza, the large square between the student union building and Sproul Hall, turned into a march, and thousands of people congregated. They tried to tear down the fence, they threw things, and they got gassed. Over the next several days, Berkeley turned into a war zone. Storefronts spit broken glass, wooden barriers and rolls of barbed wire lined the sidewalks, and there were National Guardsmen and Alameda County Sheriffs on every corner of south Berkeley. People were challenged, arrested, and beaten up. And word went out that the cops were using live ammunition! Most of the news came through word of mouth. Four hundred people were taken to a compound somewhere and made to lie in the sun on their stomachs all day long. Those who resisted were beaten. Dozens were in hospital. Someone had died. Someone had been blinded. Flowers were planted each night and the guardsmen were ripping them out each morning.

The rumours of violence were hard to refute. A friend of a friend had a black eye. We knew people who had been arrested just for opening their mouths, for asking the cops what they thought they were doing here. But my occasional trips downtown revealed only a boarded up no-man's-land, unnaturally quiet. Curfews were in effect day and night. The university was closed. Classes were cancelled. No, it seemed certain that the streets were not safe, and I mostly hunkered down with Thomas, smoking hash and gossiping on the phone.

One day, we heard that a massive demonstration was about to begin in Sproul Plaza. This was one event I didn't want to miss. I had no strong feelings about the park or the university—the cops and guardsmen scared me, but the fundamentals of the dispute seemed a bit silly. I just wanted some action. I wanted to watch a huge demonstration that might disintegrate into a riot. I wanted to see it with my own eyes, to be caught in the swell of it. And so I stood there in the crowd while speeches were made. I stood with hundreds of others on the steps of the student union. And right across from us, on the steps of Sproul Hall, was a phalanx of Blue Meanies (as we called them)—cops or guardsmen with blue uniforms, shields, and gas masks. They looked like aliens with their black, elongated snouts. Hostile aliens.

It seemed so funny: the crowd would advance, a metre or so at a time, then another metre, shout a few slogans, and then, cautiously, move still closer. And suddenly the line of Blue Meanies would step forward ten paces or so, in synch, at least two hundred of them. And everyone would scurry back to their original places, alarmed. Funny, and silly, and good entertainment on a sunny afternoon. Until the helicopters arrived. I'd never seen anything like it: the Blue Meanies now advanced continuously into the crowds. With their guns aimed at us! While the helicopters showered pepper gas down from the sky. It seemed nobody could escape. We ran in all directions, but the exits from the square were blocked by soldiers. The gas kept coming down, and people were truly terrified. They were down on their knees, coughing, choking, vomiting. I ran down into a small ravine, jumped over Strawberry Creek, and somehow kept running till it was all behind me. But I was shaken. The world appeared to be even more dangerous than I'd imagined. How could the government treat people like this? If anything ever came close to politicizing me, it was that afternoon.

I first met the Heap a few days after the Sproul Plaza showdown. The afternoon was sunny and hot, and I decided to gobble down two tabs of acid—white pills painted sky blue on one side. A good amount, I thought, to get some perspective on the strange events befalling my beloved Berkeley. An hour later, the wave of sensory information began to crest, as always, but soon I became claustrophobic. I had been inside for much of the week. Against Thomas's good advice, I decided to go for a walk.

"But Marse, there are cops everywhere! You'll get beaten up. They're in no mood to reason with acid freaks like you."

"I'll be careful. Don't worry about me."

"And we have visitors coming. I told you about those three people

John is bringing with him. I think they're sort of bisexual or something. Very cool people. You'll want to meet them."

"Thomas," I said, "I will come back in one piece, and I will meet these fine people of yours."

I walked toward Telegraph, toward the boarded-up storefronts. I was exultant. The acid was like a tonic. With my serotonin receptors now fully occupied by LSD, the mantle of caution that had sat on my nervous system lifted like a fog bank in the sun. Unfiltered stimulation poured into my sensory channels. The air was alive. It shimmered and reverberated like the ocean of molecules it was. The heat rose in coloured convection currents. I could see the tiny currents of air surrounding each flower, each fence post, like so many outlines in a badly penned drawing. My gut was full of energy and my face was bursting in a nonstop grin.

By the time I got within a block of Telegraph, I started to slow down. The place really was dead. Nobody else was on the street. On such a nice day. Fragmented associations collected in my orbitofrontal cortex. Empty streets—angry cops—broad daylight—empty streets. Cops vs. hippies. These components merged into a single, unnerving interpretation, relayed back to my amygdala by bundles of axons. My amygdala activated norepinephrine centres in my brain stem, and they quickly tuned every neuron to the spectre of threat. But what was threatening here? Not the outlines and the rainbow hues, not the hieroglyphics on the sidewalk. Those were my friends. What was threatening was the car travelling up Dwight Way just behind me, slowing down, now two or three metres away, moving at exactly my speed. A white car: a police car. My amygdala burst free of its orbitofrontal constraints. *Cop car: serious danger!* My amygdala mounted a screeching siren wail. My muscles began to tighten. Yet I could still think. Despite the serotonin drought and the tidal waves of norepinephrine. Don't look at them. Keep walking. Look casual. Don't change your pace. I kept walking. Very gradually, the car pulled up so that its windows were precisely parallel to my head. *Keep walking. Don't change your pace.*

I had to look. It would have seemed unnatural not to look. It would have seemed like I was trying to avoid them, ignore them. Hi, guys. I would be polite. My glance would be a sort of greeting, a signal from me to them, crossing from my world to theirs, a message of calm and reconciliation, belying all this suspicion and mistrust.

But when I looked over at the car, the scene was not a nice one. I saw four very large men. Two in the front and two in the back. They were wearing beige uniforms. They had guns. They were sweating. The big fat pores on their big fat faces, sitting atop their big fat bodies, were oozing streams of thick, congealing sweat. That's the image that held my panicked brain. And they were looking at me. All of them. Their faces were masks of slavering hostility. The dripping sweat and the dripping hostility, and the little round eyes, and the gross creases between their bulbous necks and chins, and the ramifying, mushrooming expansion of their huge fat heads, and of course the indescribably powerful shotguns sitting one in front of each man. I was very scared.

"Hey, kid."

"Yes, sir."

"Whatcha doin'?"

"Just walkin'."

"Where you walkin' to?"

"Just going for a walk."

"Are you one of them hippie faggots?"

No answer.

"I asked you if you're one of them hippie faggots?"

"No, sir."

"Then how come you look like one?" Laughter.

And they just kept cruising right next to me, staring at me, with evil little smiles on their suppurating faces. I slowed down. Nobody can drive this slow. But they did. Any moment now, they're going to get out of the car and beat the shit out of me. I tried to decide, with the tiny part of my brain not yet paralyzed by norepinephrine: Which way should I run? Or should I just let them beat me? And then: The best thing to do is turn around. Dwight is a one-way street.

"Well, gotta go home now," I said politely, and I slowly—very slowly so that they wouldn't notice—turned around and started walking the other way. They would have had to drive against traffic— what traffic?—if they were going to follow me.

I walked. I noticed no car beside me. No squealing brakes. No shouts of indignation. No catcalls. No amplified commands to lie down on the sidewalk. No gunshots. In a little while, I felt a little safer, and I walked a little faster. And then I ran all the way back to the apartment.

In just over a year I'd been busted five times, spent two nights in jail, and generally learned to exercise caution when dealing with officers of the law. But I'd never been terrified of the cops till now. When I burst into the door of our apartment, I was breathing hard with panic. I was certain I'd come very close to having my face bashed in. Those guys looked like they could do it—with pleasure. The image of being pummelled by four large cops would not go away. My amygdala was still on full throttle. And of course I was on acid, which seemed to be the norm when nasty things happened to me. Serotonin is a particularly important balm for the amygdala, and without it we are prone to anxiety attacks and depression. A serotonin blanket would have been more than welcome just now, but my nervous system kept throwing it off.

And we had company. I tried to calm down enough to greet them. Our poet friend John was sitting in a corner, fastidiously rolling a joint. And spread out on our carpet in poses of pristine serenity were three people I'd never seen before. Two men and one woman, all remarkably beautiful.

I was introduced around. "This is Ralph." He looked at me with a long, inquisitive gaze. He looked into me, past the surface, absorbing truths about me—truths I couldn't imagine. His features were incredibly fine: sculpted, Mediterranean, Maltese. His black curly hair formed a dishevelled halo. And he took my hand. He didn't shake it; he just held it.

"Hello, Marc," he said softly, his eyes shining with sympathy and something else. Perhaps irony. And then, after a pause, as if taking my pulse, "It's going to be okay." His voice seemed all-knowing, unnervingly so, and I was entranced. What's going to be okay? How do you know?

Ralph was older than the rest of us. About thirty. He introduced me to the others: "This is Pumpkin." And then, seeing how I stared at her, "Isn't she beautiful?"

She looked nothing like a pumpkin. She was about my age, slender, petite, wearing the soft, loose clothing of the day, underplaying small, perfect breasts. She was like an elfin goddess, shimmering in orange and gold. I later found out that her propensity to wear orange had inspired her nickname. And yes, she was very beautiful. The symmetry of her features was crowned by a rich smile, on its way to laughter. Her eyes were green, slightly touched by pain, but wide and welcoming. She reached her hand toward me, but instead of taking my hand, she took my shoulder and pulled me toward her. Before I knew it, she was holding me. "It's bad out there, isn't it?" she crooned. I began to let myself go.

The third member of the trio was Jim, ordinary-looking compared to the others. But he was a handsome man. He could have been a model for a line of camping goods. Rugged, boyish, wholesome, and glowing with a simple confidence. Jim looked as though he was happy. Just happy. All the time.

I nodded and tried to smile. I looked at each of them. I couldn't help staring. I knew, without being told, that these three were lovers. Yes, of course: *the Heap*.

And then I started to unload, moaning and chattering.

"They were going to beat the shit out of me. Four of them!"

"But it's okay now, isn't it?" Ralph and Thomas seemed to say this in unison.

"But they looked like monsters! They were so full of hate!"

"But now you're safe with us."

"But I can't forget what they looked like. It won't go away . . ."

After a few minutes of such spewing, the others seemed to huddle, to discuss the situation. Then Ralph turned to me.

"We can help." He pulled a pill bottle out of somewhere. "It's PCP, and it's pure. Would you like to take some?"

"What will it do . . . to me . . . on acid?"

"It will make you very, very relaxed, and you won't care about the cops anymore. It will all go away."

Phencyclidine (PCP) is a powerful, tranquilizing intoxicant, which slows you down but also distorts reality with a potency that rivals that of psychedelic drugs. Yet in a completely different direction. Its action is very similar to that of dextromethorphan. PCP is an NMDA antagonist, like dextromethorphan and ketamine. A dissociative. It blocks receptors—intake valves—on the surfaces of neurons all over the brain, so that the excitatory neurotransmitter, glutamate, cannot occupy them. Recall that the NMDA receptor is a primary target for glutamate, and its main job is to speed up the firing rate of the cell it belongs to. But NMDA receptors have a special function. Like community leaders that organize neighbourhoods, they synchronize cortical neurons into networks of *coincidence detection*, ensembles that lock in "reality" through their coordinated action. They allow the cortex to process the real world, its features assembling themselves into a logical, recognizable configuration, an architecture of *sense*. And if this synchronized network gets interrupted by an NMDA antagonist, sense is lost, and meaning—personal relevance—bubbles up raw, unfettered, unlicensed, from the limbic system—the hippocampus and the amygdala. So that the user falls back into an echoey fantasy of what the world *might* be like, completely made-up from bits of memory and private associations. Strange indeed, the effects of these drugs. While ketamine is thought to simulate schizophrenia, it has also been used to treat depression. Maybe it frees the amygdala from the grip of *real* dangers. Maybe some people, sometimes, just need a break from reality.

That might be what I need, sitting with three unlikely angels, overcome by anxiety in a city savaged by state-sponsored terror. I have never tried PCP, and its reputation within Berkeley drug circles is on the shady side. But I want to trust these people—I *do* trust them, more than I trust myself. So I take the two pills Ralph holds out and down them with a glass of water, supplied by Thomas, thus mixing two of the strongest psychotropic drugs in existence. And after half an hour of skittering nerves, the PCP begins to act.

Acid opens everything up, so that even the tiny details of perception mushroom, layer by layer, an elaborate mandala. The world *out there* becomes a galaxy of sparkling connections because there are no serotonin filters modifying the stream of input. And then along comes PCP, which closes everything down, shutting off the coordinated activity that holds the outside world together in our brains, dispersing its components in a senseless tangle. So what happens to the signal-to-noise ratio when you turn down so much of the signal? You get a lot of noise. And today that noise is spiked with the unfiltered, flagrant intensity of serotonin depletion. It may be that the limbic-cortical loop becomes a closed feedback circuit that increasingly amplifies its own noise. Perceptual input, reflection, motor rehearsal all shut down. It doesn't really matter what's going on in the world or how you might choose to deal with it. In direct opposition to the urgings of LSD, it's what's inside that counts. This self-amplifying representation of the *inner* world, the instinctual world, is where the detail now ramifies. Every internally generated image becomes augmented, stylized with lurid significance, the significance of dreams. Thanks to NMDA blockage, there is no new information. That's why the loop is increasingly closed. It's the meanings already inside me that cascade to consciousness. And their default subject matter is *me*. Everyone here is here for me. Every word uttered, every gesture, every glance, by Thomas, by Ralph, by Jim, John, or Pumpkin, is intended for me. While the profundity of acid resides in the beauty of the world, the profundity of PCP resides in the embellishment of the self: the emotional world and the god-like figures that occupy it. This is as close to Freud's *id* as one could hope to get. But it's an id that's now ten storeys high and neon-lit.

The gorgeous patterning of the carpet thickens, surges, and pulsates with an inner rhythm. The PCP breaks the acid world into great flows of warm liquid, coalescing in pools of personal meaning. There is a deafening roar in my head. I can hardly sit up. My thoughts fold in on themselves. I can sense my mental muscles stiffen and atrophy. But my emotional world extends outward, a sea of tidal feelings that connects me with everyone in this room. I cannot believe how cozy I feel with Thomas. How trusting with Ralph. How sexual with Pumpkin. Impulses to touch, to possess, to be possessed are taboos turned inside out. Ralph is so beautiful, Tom is so gentle. They all love me. They know me, intimately. And I love each one of them.

Now I feel her hands. Massaging me deeply, so as to calm me and soothe me like a child. And I fall back against these hands that know me so well, these hands that are touching every cell in my body. I let my breath out and let the hands in, under my skin, to caress the core of me. I've been waiting for her all these years. And now she's finally here, to be with me. And I want her. I've never wanted anyone like this before.

Through the crystalline finery of acid, encrusted around the resonant throbbing of PCP, I lose it completely. I am crying. Ralph is rubbing my shoulders. Jim is holding my feet. Pumpkin is murmuring in my ear while she caresses me. They are my companions from a former life. From the beginning of time. I've long forgotten about the police. I want to stay with the Heap forever.

8
HEROIN, THE HEAP,
AND THE SLEEP OF THE DEAD

At the age of eighteen, the good and bad things in my life were balanced like a spring-loaded teeter-totter. The good things included Thomas and his kooky friends, Susan and the astonishing pleasures of sex she bestowed, and of course the Heap. They included California sunshine, pickup Frisbee in the park, university classes that exposed me to new ideas and intellectual adventures. And concerts and jam sessions that induced a growing fascination with music. Good things included the exotic flavours of psychedelic trips. No doubt my attachment to drugs had become way too strong, but there were moments of sublime beauty. Watching the trees of the Berkeley Hills from a nearby peak, their branches interlaced in a carpet of green liquid life. These were perceptual glories that repainted the world with extravagant hues. A world that had gone grey over two years at Tabor. And with drugs, or sometimes without them, came novel feelings of connection, not just with Susan, or with the Heap, but with the earth. Letting the limits of my self mingle with the trees, the wind, the land, and the ocean.

Then came the bad things, and they included, more than anything else, a persistent loneliness. The violence I'd experienced in the fall and my recent taste of police brutality mirrored and reinforced an internal violence, harsh feelings gathered at the core of my identity. I did not understand the substance of this turbulence or the reason for its existence. Only occasionally could I distinguish the words of my internal dialogue. Condemnations: of me, by me. For being needy, greedy, weak, and unlikeable. Most of the time, I just knew that I felt . . . wrong. I hated being alone. I could not relax. My attention was pulled by opportunities to escape, to find bigger and better roller coasters, to leave the ground and go hurtling into space. Restlessness gurgled up from some internal rupture every time the sun went down and left me by myself. When Susan was away. Or busy. When Thomas was at Dianne's house. When the Heap were engaged in whatever it was they did. When there was only me.

The bad things also included ominous rumblings from my parents' home in San Francisco. Something serious was going on. But, typical of their Toronto credo, nothing was revealed. Whatever war was being fought in that house was inaudible and invisible. We didn't, any of us, know what we were feeling. I would travel to San Francisco now and then for a home-cooked meal, but the silence at the table turned the food ashen. Silence at the table, silence in the living room, in the halls, behind closed doors, and silence in the morning when the doors opened again. I was usually glad to get back to Berkeley, where the confusion of my life at least assumed robust and exotic hues. Dark but radiant hues, as I churned my nights into froth with new and more powerful drugs.

I went to visit the Heap as often as I could. I felt as though I were their pupil, their disciple, even their pet. Sometimes they would shoo me out, but often they would let me hang around, eat with them, talk with them, lie in bed with them and listen to the soft, moaning music they liked. Ralph was kind to me but also stern, like a concerned father. He would try to give me lessons in self-improvement—to calm down, to like myself more—by describing my problems in allegorical images I half understood. "Marc, the only way to find what you are looking for is to stop looking so hard." Pumpkin was always sweet to me, but I was rarely alone with her. I don't remember spending much time with Jim, either. I liked him very much, and I suppose I envied him. I was never invited to join the Heap in their sexual activities, and I never completely understood them. Sometimes all three would spend the night together in the same big bed. Sometimes it would be Ralph and Pumpkin, sometimes Jim and Pumpkin. Sometimes Ralph and Jim. I thought that Ralph and Pumpkin were the primary couple, the king and queen, yet there was no hint of hierarchy or territory. They were surprisingly discreet. Once I spent a whole night lying on the floor beside the bed in which all three of them cuddled. Despite my curiosity, nothing else happened that night. But I was content to remain ignorant. After all, I was just their puppy, lapping up the leftover affection that drizzled from their table.

They lived in a house on the north side of Berkeley, some distance from our apartment. It was a fairly large one-storey house, with a central corridor running from the front door to a private suite of rooms at the back. The first few rooms off the corridor on either side seemed the domain of the Heap. The rooms at the back were occupied by someone named "No-no." It took me a while to understand this name. No-no would appear suddenly, at any hour, in a bathrobe or a gown of some sort. Never in normal clothes. No-no's figure was generous but not fat, topped by a soft whiskerless face and long silky black hair. No-no spoke in a contralto voice. Contralto, not tenor, because it was a womanly voice. But No-no was not a woman. At least I was almost sure he wasn't. Yet it was impossible to see him as a man either. He/she was sexless—or perhaps sex*full*. And of course that made me quite nervous. No-no would greet me with a purring, almost liquid voice, a sort of "come hither" lurking behind every utterance. Which produced in me the strongest urge to back away and yet not leave. *No-no*. His name reflected some velvet prohibition that I couldn't quite fathom.

You could say it was a house of weirdos. But for me it was a house of almost supernatural warmth, a secret mysterious mix of love and sex and hidden knowledge that attracted me, frightened me, and held me. It was a house in which I could never belong, in which I would always be a visitor, if not a stranger. And yet I felt cared for there—by Ralph and Pumpkin especially—in a way that reached to the core of my loneliness. And then it was time to go, when the Heap had other things to do.

One night, a man named Ben came to call. I had seen him before. I knew he was the Heap's drug supplier, and one of the drugs he supplied was heroin. Ben greeted me with a thin smile. Was he offering? I did not hesitate to get in line.

The Heap had a dark side, and I was starting to get it. Ralph had been a serious junkie. That means physically addicted, daily use, and all the catastrophic outcomes that go with it. I didn't know when, or for how long. He didn't seem the type. But there was something about his fragility, his delicacy, that hinted at a long-standing struggle to remain intact. As though his life were suspended above some inevitable tragedy, waiting to engulf him. Perhaps the Heap and its love were his treatment program, his support system. Perhaps they existed for the sole purpose of keeping Ralph from the jagged rocks of relapse. I sometimes imagined that. Yet it began to seem that Ralph relapsed often. I was never fully aware of how frequently Ralph shot smack, or whether this sporadic use should be considered a complete fall from grace or just . . . a weakness. He didn't discuss it with me, nor did the others, but no one hid it from me either. In the Heap's strange credo of openness, heroin was allowed to exist, as a malignant ghost, and for me the most potent of attractions.

I saw heroin as a badge of courage and a rite of passage to the inner chamber. I didn't want to become a junkie, but I did want to try it a few times. If I could not join the Heap in sex, if I could not stay with them day in and day out, then at least I could follow them, part-way, along this dark path.

About half an hour after Ben arrived, I passed Ralph in the hall. He was coming back from the bathroom, walking heavily, his eyes hooded. He glanced up at me for one brief moment. He knew, and I knew he knew, exactly what my intentions were. His gaze was somehow scornful, sympathetic, and terribly sad at the same time. Why didn't he stop me? I looked away and kept walking. I entered the large, brightly lit bathroom. Ben had just shot himself up. He seemed very relaxed but not spaced out. He was an accomplished junkie.

"Do you want to try it, Little Marc?" For some reason they called me that.

"Yes. How much will it cost?"

"Is it okay with Ralph? Don't you think we should ask him?"

"Ralph saw me come in. He knows. It's okay." The last thing I wanted then was to ask permission. I just wanted to *do it*. To cross over. To lose this final virginity.

"I'd say seven bucks' worth for a lightweight like you."

I knew that heroin was evil. Its reputation was unambiguous. What would Thomas think? He'd say something like "You're going to shoot smack? Heroin? Why don't you just lie down in front of a truck?" But I didn't expect to like it that much. Wasn't it sort of a downer? I wasn't attracted to drugs that made you feel dazed and sleepy. And that needle stuff was creepy. I watched Ben prepare it. He had the needle in his fingertips. A small metallic barb attached to what looked like an eye-dropper. He drew some liquid into the dropper from a soup spoon, after holding it over a match flame. The liquid was light brown in colour. He pulled a belt tightly around my arm and asked me to hold it. And then I felt the pinprick in my arm. I saw a spot of blood in the needle itself, and then I looked away, because the sight of it was repulsive and frightening. I had no idea what was charging into my body through that thin metal passage.

I begin the familiar routine of waiting. I am accustomed to waiting for drugs to take effect. But this time there is no waiting. Within ten seconds the world begins to change. The things around me, the walls of the too-bright bathroom, the air itself, are suddenly spinning. And the centre of this vortex is me. I am the funnel of a spinning world, sinking down, down, into the floor. Disappearing, but at the same time coalescing. So very much here. I am completely present, completely still, yet falling. The weight of my body is enormous. A condensed mass. But this weight is not a physical thing: it's a nexus of bodily comfort and emotional wellbeing. A warm syrup. There is no sleepiness, no drowsiness. Though I stagger and have to sit down, there's a kind of midnight clarity in which weight and weightlessness coincide. And I am the centre of it. Outside of me nothing exists. Ben is somewhere far away, at the other end of the room. His grinning face peers at me from elsewhere.

The natural opioids—those manufactured in our own brains— are produced in the hypothalamus, which sits atop the brain stem, a chemical command centre. They travel to many other parts of the brain, to the spinal cord, and to parts of the body like the stomach and skin. Opioids make up one branch of the chemical family known as neuropeptides, a family that has earned the name "molecules of emotion." Peptides tell you what's going on in your emotional world. In large part, they *are* your emotional world. They make you feel much of what you feel when you're angry, or excited, or in love. Opioids are among the most important of the neuropeptides, and our brains would not be able to function without their daily support.

But the molecules now coursing through my neural passages didn't come from my hypothalamus; they came from some drug factory in Mexico. And it's no wonder that humans have learned to manufacture molecules that do the same job as natural opioids—considering the benefits they bestow.

Natural opioids include "endorphins," known for the high that runners get after a nice long jog in the park. But opioids have much more serious business to attend to. The primary functions of the opioids are threefold: to provide relief from pain or stress, to produce a sense of pleasure or wellbeing that can energize any goal, and to use either or both of these feelings—relief and/or reward—as the emotional currency of human attachment. Mother's milk contains opioids, which seems a kind of original sin when it comes to drug addiction. Babies love opioids, and presumably their mothers, because of the feelings of warmth and safety produced by these molecules. Natural opioids are released in response to the soothing touch of a parent or lover, to cuddling and holding, and even to play. But they are also produced in humans and many other species in response to pain, fear, and trauma—either physical or emotional. When we are wounded, whether by anxiety, loneliness, or injury, it is the opioids that tend our wounds. But how can this be? How can very good and very bad things—a mother's embrace and a torn limb—both produce the same effect? It seems that the *feel-good* function of opioids evolved first for pain relief and then got borrowed for other purposes. Such is the economy of evolution: supplies are limited, so make the most of them. But here's another mystery. Opioids make you feel safe, warm, and capable—is that pain relief, or is it pleasure? Perhaps a sense of relief is the main ingredient in the mammalian formula for feeling good. One way or another, natural selection found a winner: pain relief and pleasure rolled into one molecule. And, while the evolution of opioids dates back 150 million years, it was just a few thousand years ago that similar molecules were discovered in the flowers of the opium poppy. Serendipity indeed.

Opioids make you feel good in at least two ways. First, by inhibiting the firing of neurons that are activated by pain or stress. These neurons are found all over the brain, in locations such as the spinal cord, where pain is first processed; the brain stem; a region of cortex called the insula, where pain and other feelings are *felt* and made conscious; the amygdala, where emotional reactions are orchestrated; and even the prefrontal cortex, where the world is evaluated and acted upon. Why do we need opioid receptors in so many places? Because there are so many levels to pain and suffering, from the physical to the mental; suffering is such a complex process, it is literally constructed all over the brain. And not only is the brain full of opioid receptors, but their density increases as you move up from the brain stem into the cortical areas responsible for conscious thoughts and feelings. That means that the aspects of suffering most susceptible to opioid relief are psychological, not sensory.

The second way opioids make you feel good is by targeting receptors in a very specific brain region: the ventral striatum. As described previously, the ventral striatum is the home of motivation, of forward thrust, wanting, or desire, once so potently activated in me by visions of Lisa waiting in New York. Desire is ignited by dopamine as it gets pumped through the ventral striatum. But desire is not the same as pleasure. According to one prominent theory, "wanting" and "liking" are distinct psychological states underpinned by distinct neural processes. It's the opioids that cause *liking*—the sensation of pleasure or wellbeing. And it's dopamine that produces *wanting*—the feeling of desire or attraction. Both neurochemicals do much of their job in the ventral striatum, where there are receptors for each. And because they are neighbours, anatomically speaking, these receptors are well positioned to work together.

The emotional circuitry of the ventral striatum seems to derive its power from an intimate discourse between opioid *liking* and dopamine *wanting*. In contrast to their calming function, opioids actually increase the flow of dopamine in the striatum, by relaxing the brakes that hold it back where it's actually manufactured—in the ventral tegmental area or VTA (see Figure 3). So, in the neural mechanics of feeling good, the excitement orchestrated by dopamine joins the soothing balm of opioids. This partnership is the basis of "reward"— the term psychologists use for whatever makes us feel good and want more. And it's this combination of opioids and dopamine that unlocks the tide of positive emotion I feel tonight in the Heap's bathroom. Yet this partnership has a more important task than conferring happiness: it gets us to work for things. It makes evolutionary sense that anything that feels good should become the target of goal-seeking. And to motivate goal-seeking, liking and wanting have to become connected. First you feel good, then you want more. At the brain level, opioids in the ventral striatum cause the feeling of wellbeing, but then they trigger dopamine release, enhancing the appeal of whatever's showing up on the screen of perception. Natural goodies like food and sex certainly follow the progression from liking to wanting. Feels good— want more. But with goodies both natural and acquired, it is dopamine's flame of desire, unleashed by the *ahhhhh* of opioids, that causes animals to repeat behaviours that lead to satisfaction. Here in one neat package is the fundamental chemistry of *learning*, which really means learning what feels good and how to get more of it. Yet there's a downside: the slippery slope, the repetition compulsion, that constitutes addiction. In other words, addiction may be a form of learning gone bad. For me, this neurochemical sleight of hand promises much more pain than pleasure in the years to come.

Heroin is many times stronger than morphine, which is many times stronger than the opioids produced by our own bodies. Yet heroin activates all the same receptor sites, from spinal cord to striatum to cortex, penetrating every level from body to mind. Those receptor sites were designed to respond to a trickle of opioid molecules, not a flood. So the throng of molecules now landing in my opioid receptors has an impact well beyond anything evolution had in mind. The high-strung neurons in my spinal cord and brain stem are first to slow their firing rates: my legs feel leaden and my arousal level comes crashing down to a sonorous background chant. The rapidly firing cells of my amygdala decelerate to a low thrum of contentment. Neurons throughout my limbic system and cortex change their rhythm, disengaging from neighbourhoods tuned to the stress of loneliness, the fear of depression, and the immediate threat of reprisal—from Ralph, from my own inner voices, and from parents who will surely condemn what I have done. I stop caring. All threat is neutralized. There is no danger anywhere. And in my striatum, opioid sponges are soaked with heroin, turning off the need to strive, turning on the glow of wellbeing, but also, insidiously, revving up the dopamine pump that tops this dark lake with an electric sheen of attraction. This moment is lit up with glory. And soon dopamine's urgings will progress from exhilaration to desire, from crowing to craving, leaving an incessant hunger in my cells.

For now, the impression of vertigo disappears, and I feel only the soothing tide of peace and relief, the stuff of mother love multiplied a thousandfold. This powerful mother puts her arms around me and crushes me with her warmth, squeezes me with her nurturance. I feel relief from that pervasive hiss of wrongness. Every emotional wound, every bruise, every ache in my psyche, the background noise of angst itself, is soaked with a balm of unbelievable potency. There is a ringing stillness. The sense of impending harm, of danger, of attack, both from within and without, is washed away. My opioid receptors, choking on enormous gulps of their one and only food, send commands to the cell bodies that host them. The activity of these cells is turned down in many, many places, diminished, almost neutralized. And in a few very special places, where pleasure is manufactured, it is turned up.

I slowly make my way along the bathroom wall, toward the door. Time becomes wholly personal. It doesn't collapse, as with LSD; it stretches like toffee, cushioning my thoughts and extruding them into pseudopods—impressions, waking dreams. There's Ralph. The face that seemed so heart-rending and tragic now seems friendly and even conspiratorial. We greet each other like characters in a cartoon. Like puppies in a litter. He is the big brother, miffed but forgiving, unruffled. I am the bad boy who somehow ended up next to him in this scene, finding him familiar in this strange world. We are caricatures of our previous selves. Our conversation is effortless but also irrelevant, a banter of superficial phrases, interspersed with long pauses in which we lose track of each other and sink instead into our private reveries. Eventually I saunter into the living room, sit down, and stare into the dreamy fog that has settled over this house. There is no way out of here. Not for now, and maybe not for a long time. And that's all right with me.

I bought heroin from Ben every couple of weeks, usually when he came to visit the Heap but a few times by private appointment. I got my own needle, so I wouldn't have to borrow someone else's and try to ignore the risks that entailed. I learned how to clean it and care for it. I learned how to find the best veins in my arms. I experimented with different veins—in my arms, my hands, even my feet— sometimes creating massive bruises when the liquid missed the vein and filled the cavity between skin and muscle, turning patches of skin purple and blue. I kept my sleeves rolled down. I loved the ritual of cooking up the smack in a spoon, drawing it into the needle through a small piece of cotton. I wouldn't let Ben do it anymore. I did it myself. And then came the relief mixed with crazed excitement when I successfully penetrated the vein, watched the red backwash of my own blood, and squeezed the liquid home to my cells. It still revolted me in some ways, but less and less each time, and soon the revulsion turned to fascination and the fascination to attraction. Yet over and above the pleasure of the high itself, I was filled with a certain pride. I was really a bad boy now, playing with the most powerful drug of all. I had been inducted into the secrets of a dark society, and now I held the keys.

I sometimes shot up with Ralph, but usually he kept his troublesome heroin habit to himself. I did it with Ben, or alone, in my own apartment, or sometimes in a lit doorway, revelling in the *wrongness* of it, mixing the thrill of norepinephrine-spiked anxiety with the other ingredients of my neural soup. Or I mixed it with other drugs. I sometimes took it after a capsule of mescaline or psilocybin, arranging the tracery of psychedelic images in the echoey dream world of opiate calm. My primary objective was to become a master chef, concocting the flavours of my internal world with new and daring recipes. Not only to switch my feeling state to one that was tolerable, as I had learned to do at Tabor, but to select a state that would best suit my mood and my means. This seemed a fine achievement. Yet behind it lay the final lap of a more profound victory: complete confidence that I could make myself feel happy and free without relying on others, without the cooperation of anyone or anything—without consent.

I continued to visit the Heap during these weeks. My new hobby did not seem to alarm them. And why should it? They managed their lives through the manipulation of their internal chemistry more than anyone I knew.

Their involvement with drugs made them more magical and more impressive in my eyes. I didn't see that the kind of drugs they used and the intimacy of their relationship with those drugs would inevitably lead to disaster.

One night, I got my first glimpse of that impending doom, watching a terrible bedtime ritual take place between Ralph and Jim. It was apparently one of the nights Ralph and Pumpkin were to be together and Jim was to sleep alone. Jim got into bed while the others were still up and about, but he wanted a special goodnight from Ralph. I watched from the doorway. They probably weren't even aware of my presence. Ralph pulled a syringe out of somewhere. Jim held out his arm. His eyes glowed with a beatific trust. Ralph asked Jim to make a fist and he did so. There was no cord or belt necessary. Ralph found a vein and pushed his needle into it almost tenderly. Some clear liquid mixed with Jim's blood in the needle, and then Jim's whole body undulated in a wave of physical relaxation. His head fell back on the pillow and his eyes closed gently. The last thing I noticed was his smile, which lingered long after his eyes had closed for the night. Ralph looked up at me as he prepared to leave the room. If he was startled by my presence, he did not let on.

"What was it, Ralph?"

"Barbiturate. Seconal. What does it matter?"

"But why . . . why does he need that?"

"It helps him to sleep." Ralph made to pass by me. He was in no mood to banter about the beauty or the ugliness of shooting someone to sleep. But he must have seen my shock and consternation.

"It's okay, Little Marc. It doesn't happen every night. It's . . . something special. Something he just . . . likes."

I forgot about Jim's midnight send-off for a while. Life went on. I continued my efforts to get as high as I could for as long as I could. And then, one night, the wave broke. It was the summer of 1969, and in this particular week I seemed to be on a mission to take as many different kinds of drugs as anyone possibly could, one after another, their effects overlapping and compiling. I started off in the house of a friend of a friend, hanging out with a couple of other guys, smoking a combination of hash and opium. There were instruments there. I played music for half the night. Then came the psychedelics: mescaline or psilocin, a synthetic form of psilocybin. Two hits for me, please. I always take two of everything. There must have been some form of speed involved, probably Benzedrine, because I was wired all night and well into the next day. By late that afternoon, I had migrated over to the Heap's place. I hung around, stoned, dazed, but still hungry. In fact, more hungry as night approached, because I had been high for so long that I couldn't bring myself to end it. I didn't want to surrender, and I didn't want to go home by myself. Beside that, an undercurrent of bodily carnage had begun to penetrate my con- sciousness. I didn't feel great, and the only way to feel better seemed to involve taking more drugs. That's when Ben showed up.

At first, Ralph pretended that Ben had come empty-handed, just to visit. But when I barged into the bathroom and found him setting up his stuff, Ralph became adamant. "Stay away from it, Little Marc, don't do it. You've got so many drugs in you already."

"But . . ."

"No buts! You haven't slept for thirty-six hours. There's no telling how you'll be affected." He had never seemed so severe.

"But, if I just take a little . . ."

"You won't be able to judge. There's no telling—"

"Yes I will! I've gotten pretty good at figuring out what I need, and—"

"Look, I'm asking you not to do this. For me."

He kept walking. In despair? In disgust? Did he see the addict in himself in me, and turn away from me? Or from the hypocrisy of saying no to me? Or did he simply recognize my feverish determination and decide to let fate take its course?

I *was* determined. There was really only one drug left to take that night. And it was right there, in the next room. I evaded Ralph half an hour later and got to the bathroom just as Ben was packing up. I had to beg. I didn't know how many people, how many of the Heap, had required his ministrations that night. He seemed to be done dispensing, but I would not take no for an answer.

"Come on, Ben. Just give it to me. I'll do it myself if you have to go." He sold me ten dollars' worth—not a huge quantity. I wanted to be careful. Ralph's admonishments stayed with me. Just not enough to make a difference.

Ben shot me up. As a favour. I thanked him. My arm was numb where he tied it off, just above the elbow. And then he was ready. And I gave myself over to him, to it, with no concern for anything but the completion of the act. I just wanted to have it in me.

My next moment of consciousness could have been hours or days later. There was no association with Ben or with heroin, at least not at first. I was in a bathtub with most of my clothes removed. Ralph, Pumpkin, and Jim were all there, looking down at me, and their faces expressed a surprising mixture of anger, relief, and affection.

"Whassa matter, guys?" My voice came out in a croak, my words mashed together. "Why . . . why am I in here?"

"You OD'd, Marc." Ralph's voice was terribly harsh. "We thought you were dead. We were just talking about what to do with the body. We were trying to figure out where to get a big enough bag." His gaze was unrelenting. He had warned me.

"Aw c'mon, I'm not dead, am I?" I slurred.

"Your heart stopped beating for several minutes," Pumpkin said.

"This is not a joke," Ralph continued. "If we hadn't dumped you in a cold bath, that would've been it."

"But . . . ish a warm bath . . ."

"No, it feels warm because your body temperature is so low. You were dead!" Ralph kept repeating this, to shock me, I suppose. To scare me. As if that could undo what had already happened. But it did not penetrate. I was very stoned. Sedated, apathetic, and unreceptive to bad news or good news. I didn't get the full picture. Was his portrayal accurate? Respiratory suppression is what underlies an opiate overdose, and cold water won't help someone's breathing. In fact it can make matters worse. My pulse must have become very weak, but it probably did not stop altogether. Yet I accepted the Heap's account for years to come, and I felt a body blow of shame whenever I imagined it.

I was taken to the house of another friend—Dave or something. Not a close friend, but a good-hearted fellow who was willing to help. I didn't know why I was taken there. I imagined it was out of disgust: the Heap wanted to be rid of me. Dave called some neighbourhood drug clinic, and they sent someone over right away. For the next hour or two, a self-righteous and pitiless young man sat across from me and abused me verbally, denigrating me for the stupid thing I had done, for the anxiety and pain I had caused my friends, for my gross irresponsibility. "Do you care about yourself so little? Do you think life is so cheap? You're willing to throw it away for a few hours' high?" His lip tightened at the word *high*, as if he could hardly speak it, and I sat there on a bench at the kitchen table, absorbing his insults. I didn't see the point of it all. Dave later told me the guy was trying to make me mad, to invoke my fighting spirit so that I wouldn't go back into a stupor again. Anger causes a wave of activation to descend from the hypothalamus, turning on the lights all over the sympathetic nervous system. And the resulting adrenalin rush could indeed kick a sleepy heart or deadened lungs back to life. Maybe it was working. But I felt little conscious anger, little of anything. Instead, the mixture of heroin and other drugs brought forth a swirling dreamlike state poised on the border of hallucination. I was told to take a walk and to keep moving, no matter what.

There was a young woman living at Dave's house: Tina—slim, dressed in leathers. I thought she was sexy. She offered to accompany me and watch over me. So we walked along the sidewalks of Berkeley, whether for a few minutes or a few hours, I couldn't tell. I was immersed in a thick, warm cloud. We were supposed to keep walking until the sedation started to wear off, but it was difficult for me to monitor any aspect of my physical state. In the dark streets, the thin seam between waking and dreaming opened into a world of fantastic images. I saw a golden chariot, embellished in ancient symbols. It was pulled through the street beside us by what appeared to be a pair of giant dogs. I grabbed Tina's hand tightly—and looked away. I'd never been this high, high enough to conjure up detailed visual hallucinations. But I was not displeased. I had achieved some peak experience that now, finally, signified the top of the mountain. I had gotten as high as it was possible to get. I did not want to go any further.

The Heap got married a few months later. Not officially, of course. I was one of thirty or so people at their wedding service, performed by some priest-like figure whose incantations seemed to borrow from many religions. The ceremony took place in the Berkeley Hills, in the tall grasses of Tilden Park, with daisies everywhere: in the grass, in Pumpkin's hair. Ralph, Pumpkin, and Jim were all dressed in pure white, and they were impossibly beautiful. It was painful to look at them. The daisies in Pumpkin's hair formed a circular braid, which seemed to me the ultimate symbol of their union. And then I neither saw them nor heard from them for months.

I didn't know why. Was there a rift in our friendship, made permanent by their anger at me? For my disobedience? For my brush with death? Or were they trying to protect me from the wash of drugs that inevitably flowed through their lives?

My overdose experience was not enough to turn me away from hard drugs for good. But it was enough to shock me into being careful. I made a promise to myself that I would stay away from smack and from needles. I recited this promise out loud one night, imbuing it with as much gravity as I could manage, urging myself to believe it. Whatever drugs came my way would be ingested through mouth, nose, or lungs—not through the vein in my arm. And even when that promise expired, more than a year later, my experiments with drugs retained some small measure of caution—for a while.

My connection to the Heap never rekindled. I saw Ralph once or twice more. And then nothing. Maybe we said goodbye, or maybe we just stopped.

Then, two years later, I learned of the terrible tragedy that had torn the Heap to pieces. Jim had overdosed on barbiturates. He had shot himself up alone somewhere, and he had died, choking on his own vomit. I don't know who told me this, but I did not doubt it. It was almost inevitable that the Heap would meet with catastrophe. They seemed, at least in my eyes, both too beautiful and too flawed to exist for long in this world. Jim had died. And what about Ralph and Pumpkin? Ralph ended up in a mental institution of some sort, so the rumours went. And then, when he got out, he and Pumpkin had married. This time officially. I did not try to find them. I could not conceive of Ralph and Pumpkin as a traditional couple—a Heap of only two. I couldn't imagine Jim dead. And I didn't want to imagine how it must have savaged Ralph's heart to lose him. Because Ralph was the daddy, the one responsible, or the closest to it. It must have been Ralph who allowed, or perhaps required, a lifestyle inundated with the most destructive drugs. I could not imagine the guilt that would occupy the third place in the triangle from then on.

The Heap was fragile. They knew they wouldn't last. But the drugs they'd relied on to relieve their anxiety, their fear of loss, turned out to be what finished them. They had nearly finished me.

GETTING DOWN

I dropped out of university in March 1970, two terms into my second year. I had done little serious work, and my grades were so low I'd been placed on academic probation, an ominous formality that meant I could be expelled if things got worse. I wasn't proud of that, nor was I inspired to work harder, so I took a break. I went to work in an insurance company as a stock boy (official title: inventory clerk). I spent my days filing forms, retrieving forms, mailing forms, and once a day delivering inter-office memos to four floors of female secretaries. Susan had faded from the picture, and one girlfriend after another took her place. Love played little part in these adventures, but I learned something about women, about how to combine sex and friendship, and, in case I needed a reminder, I learned that I didn't like being alone. I bought a motorcycle, a Triumph 500, that I constantly repaired, polished, and loved. I practised my guitar. When summer came, Michael and I played and sang on street corners in San Francisco, for the joy of it more than the few dollars the tourists threw into our hats. Then, if I didn't have anyone to spend the night with, I'd ride back over the Bay Bridge to my rented room in Oakland.

My parents had been separated for more than a year. It seemed as if their marriage had developed an allergic reaction to the California air, to the pure, bright honesty of the place. Mom introduced me to her new boyfriend a few months after the big event. Now they were living together in Berkeley, tucked into a stucco-walled apartment, gazing inward at the marvels of their relationship. Dad had rented a large bachelor apartment on Broderick Street, in the middle of San Francisco. It was spacious and bright, but its most distinctive feature in my eyes was the view from its windows: rolling fog by day, haloed streetlights by night, hinting at uncharted mysteries in this ethereal city.

After he got over the shock of Mom's absence and the sting of being replaced by another man, Dad started going out on dates. Michael and I were proud of him: he brought home women — one for a few months, then another for a few more months, then another. Where had he learned to do that? These women always seemed attractive, warm, and fun to be with. Michael and I approved. I came to San Francisco to visit whenever I could, sometimes with a girl-friend, sometimes single and hungry for company. On weekend mornings the three of us — us guys — would joke about our exploits around the breakfast table. We were learning to be cool.

One room of the apartment had tie-dyed fabrics covering every wall, a curtain of beads for a door, a drooping cloth canopy suspended from the ceiling, and an ornate copper lamp hanging from its centre. The only furniture in the room was a small wooden table, right under the lamp, sporting a very large hookah, with dozens of coloured cushions surrounding it. We called it the Indian Room, and there we smoked pot together on the occasional nights our paths crossed. When we ran out of things to talk about (which didn't take long), we'd each gather up a little pile of cushions and lean back, nestling dreamily, listening to Ravi Shankar's sitar music emanating from speakers hidden behind the fabric walls. That music got under my skin, more than I knew. It foreshadowed a journey that would eventually take me to the winding streets of Calcutta. But for now it was background to our rambling ruminations.

Michael had moved into the apartment with Dad and stayed about a year. But when September came, he shook himself loose and moved back to Toronto, where he found a guru and began to practise strange metaphysical rituals that turned his skin pasty white and his conversation to gibberish. About that time, Dad met Connie, someone he finally loved enough to stay with. She had grown up in Big Sur, got expelled from Catholic school, and now seemed to know half the cool people in San Francisco. We liked her. In a few more months, they announced their plan to get married. They were about to merge two dysfunctional families into one.

In November I quit my job and started truck-driving lessons. I'd been a stock boy for ten months, and it had finally worn me down. Now I rode my motorcycle over to San Francisco in the afternoons and practised driving a semi-trailer around the massive parking lot surrounding the baseball stadium. Three black men in their thirties, Ike, Hayes, and Lake—IHL Trucking—taught me how to shift through thirteen forward gears. They did it for nothing, out of kindness, and whenever I missed a gear I'd inevitably get chided for being a useless honky. I wanted them to like me—as much as I liked them. And I wanted to live in San Francisco. My room in Oakland got increasingly lonely without my supply of secretaries. So I moved my stuff over to Broderick Street, gradually taking over the room that Michael had left.

You could almost hear the silence in that place. White pile carpet flowing seamlessly from room to room, connecting them like tide pools. At night I would wander back and forth, hypnotized by this featureless surface. Dad usually stayed at Connie's house. I told myself I didn't mind—I had the whole place to myself. I would fall asleep in a pile of cushions in the Indian room, saturated with smoke, booze, or something stronger.

I finally got Hayes to take me to the licensing bureau to get my truck-driving learner's permit. But the truck licensing guy was out sick that day, and I never got Ike, Hayes, or Lake to take another morning off work. By the time I turned twenty, in January of '71, I'd pretty much given up on being a truck driver. I was completely unemployed and uninspired.

In the past year I'd flirted with all kinds of drugs, but I'd stayed away from heroin. Now the thought of it wormed into my consciousness from below. Living in San Francisco was supposed to have been a fabulous new beginning, a chance to make friends and meet women. I'd been envious of Michael and Dad's easy lifestyle. I imagined a closeness that I hadn't been able to find, with Dad or anyone else. I'd wanted to join the party, and now here I was — but there was no party. The nights were long and empty. The apartment was just a big, hollow gallery, resonating with whispers of good times, images of women I'd been close to, thoughts of drugs and the power they bestowed. One evening, after a tormented hour of bargaining with myself, with darkness coalescing outside and no impetus to turn on the lights, the momentum of drug images reached some threshold: my bad half (as I now saw it) won out. I decided I was going to shoot smack again. Once I accepted the decision, I knew it had been building for days, maybe weeks, without reflection. I phoned around, one acquaintance led to another, and I finally got the name of someone named Jimmy — a young guy living nearby.

When I met him the next day, I thought I might have found a new friend. Jimmy was tall, with wavy black hair, handsome, with a smile that let you right in, and a confident, even reckless style of self-revelation. He was lighthearted and boyish; we had a good feeling for each other. We were hanging out in his apartment, sharing stories, when I asked him point blank if he had any smack connections. He blithely told me that he'd been a junkie until just recently. I didn't know what to make of that. He was an upper-middle-class kid, like me. I tried not to act surprised.

"But now I'm off it," he declared. We were sprawled on old wooden chairs around his kitchen table, his roommates poking around, coming and going, grabbing beers from the fridge.

"For good?" I asked.

"Yup, for good. It's been six weeks. And I'm not going back."

I nodded, nonchalantly, coolly, I thought, hiding my disappointment. Too bad I hadn't met him six weeks ago. It would be fun to get high with a guy like Jimmy. But . . . what was I thinking? He had been a junkie, an addict. The word *junkie* still stank of sickness, of failure.

An hour later, one of us brought up the subject of heroin again. Probably Jimmy, boasting about his exploits, his connections on the Fillmore, where the only white guys were big and muscled, and even they disappeared at dusk. Fillmore was the grand avenue of the ghetto, nestled between San Francisco's wealthier neighbourhoods. The thought of Jimmy scoring on the Fillmore stuck in my head. I took a breath and asked him, "If you . . . got some, for me, would it . . . you know, would you be in danger of going back to it?"

"Nah. No way. Your wish is my command." I thought he was joking, or boasting. I couldn't detect any shift in his emotions, but my own heartbeat shifted gears.

Dad had gone to Switzerland on a ski vacation with a couple of doctor friends: a last hurrah before his upcoming wedding. He had left me the keys to his Alfa Romeo, and it was parked down the street. Jimmy walked around the car, whistled, and got in the passenger seat.

Soon we were cruising down Fillmore Avenue. It was still light enough to see, but darkness was on its way. Grim buildings gaped at us as we drove slowly by. A ragged procession of wounded-looking people drifted along the sidewalks.

"Who are we looking for?" I asked.

"I don't know his name . . . I just know what he looks like."

"You sure I need thirty dollars' worth? It seems like . . . a lot."

"No harm in having some left over. For tomorrow."

"Good idea." I peered across at him. He sat there placidly, scanning the sidewalk. Was he going to all this trouble for me? That seemed more and more unlikely. Should I tell him to forget it? That would be the right thing to do. The moral thing. Sure, I was breaking my promise, deciding to do smack again. But that didn't make me completely unethical.

"Slow down!" he yelled, then, "Stop!" I pulled over, my ruminations scattering. Jimmy jumped out of the car without another word. He was gone for about ten minutes. And then his face popped up at my window, grinning crazily. Elation surged through me: it was really going to happen. Today. Soon.

We went back to Dad's apartment. I parked the car and led Jimmy upstairs. He walked around the place, nodding his approval. He looked innocently happy, excited, like a kid. I got him a drink and he sat down in the one recliner in the living room. I waited. He put his feet up and leaned back, the picture of nonchalance. I kept waiting. And then, finally, he pulled three little balls out of his pocket, looked over at me, and winked.

"But Jimmy, you're not going to . . . do it yourself. Are you?" By now I knew the answer. I was just going through the motions. "I wouldn't want you to . . . I wouldn't want to be responsible."

"Hey, it's not like I'm going back to being a junkie, you know. I can still get high once in a while. That doesn't make you a junkie."

I began to see Jimmy in a new way. There was something intrinsically false about him. He played it the way he wanted it to be. He couldn't care less about the disjunct between what he said now and what he'd said an hour ago. I would eventually come to recognize this as a distinguishing feature of the species *junkie generis*. This creature had an amazing capacity to reinvent reality, to change its shape at will.

"Let's do it in the kitchen," he said. Slowly he unwrapped the first of the three balls over the counter. They were balloons. Airless balloons. Once he'd pulled it free of its tangled folds, the first balloon revealed a bulge at one end. Jimmy stretched it out, holding the bulge downward, then snapped the tied opening against it a few times, loosening the powder that clung to the sides. I brought him scissors. He was serious now. No more smiles and chuckles. He cut the balloon above the bulge and poured a small mound of dirt-brown powder onto the white countertop. He rolled the balloon end between his fingers, emptying the last few crumbs. Next, he slipped the edge of a piece of paper under half the mound, and raised the powder to his eyes. He looked satisfied. Out came the needle—a real syringe, not an eye-dropper. I brought him a tablespoon. He carefully tapped the mound of powder into it, then asked me to light the gas stove. In a few moments, brown liquid bubbled up from the heated spoon. He brought out a pack of cigarettes and pulled the filter off one. Then he tore off a piece of it and dropped it into the boiling liquid.

"Good idea," I said, mindlessly, mesmerized.

"You go first," he replied. I took a freshly packaged needle from him and unwrapped it, drew up the fluid, and found the large vein in the crook of my right arm, just as I remembered. Like riding a bicycle. I stabbed delicately and hit home. My heart was racing. But I wanted to go slowly. I pictured the scene in the Heap's bathroom. I didn't want to die. Once my blood appeared in the syringe, I pushed the plunger bit by bit, waiting to feel it, not wanting to be overwhelmed, but wanting, now badly wanting, to *feel* it.

It caught me in the gut first. A sudden metallic hollowness as my abdominal muscles began to lock. Not at all unpleasant, but shocking in its intensity. Then it hit my legs. I felt them go leaden. I pulled the needle out.

"Enough," I said, my voice already going hoarse. Or was it too much? That unmistakable swell caught me: I was receding into myself, contracting in a rich, warm syrup of sedation. But I didn't lose it. My consciousness was indrawn, a kernel of its normal span, but I was intact. Very intact now. The feeling of it, after so long, produced a sunburst of pure joy that I had not expected. This was very good indeed. I smiled up at Jimmy and cruised slowly into the living room, like a ship loaded with cargo, low in the water. The recliner was mine now. I spent the next few hours sitting, comfortably, just sitting, watching, dreaming, chatting with Jimmy across the room, close yet miles away, checking the weight in my belly and my legs to make sure it was still as great as before. When the feelings began to slacken, I made my way back to the kitchen. Jimmy had finished the first balloon and half the second. I poured the rest of the second balloon into a new spoon and started the ritual once again. I peered into the living room. Jimmy was sitting, humming gently. I thought he was a good friend, after all.

Jimmy didn't go home. We just lapsed into sleep sometime after midnight. I sent him to my dad's bed and slept in Michael's. The next day we woke up around noon. We hung around for a few hours. There was nothing to do, really, except finish off the smack.

"But it won't be enough for both of us." Jimmy looked at me without warmth. Did he expect me to hand it over? I wanted to get good and high one more time—at least as high as last night. We decided to pool our resources and score some more. First we loaded up with supplies: eggs, bacon, bread, milk, chocolate bars, cookies, potato chips. Then we drove back to the Fillmore and waited around, for nearly two hours, until Mr. Nobody showed up. We got back to the apartment and shot up. The night went by much as before. By the next day, it was as if we'd become roommates. Nothing official, but Jimmy clearly wasn't going anywhere as long as we were on a roll. We dropped by his house to get a change of clothes, and then we were off to score again. The Fillmore connection was nowhere to be found, but Jimmy knew someone in North Beach. We scored, we went home, and the pattern repeated itself.

When did Jimmy's money run out? By Day 3 or 4 it was pretty much up to me. He had schemes, plans, ways he was going to pay me back. But I was the one who inevitably went to the bank. Then *my* money ran out. But I still didn't want it to stop. Not yet. I really wanted to do it again. I couldn't identify what it was that felt so dreary about the idea of stopping. It wasn't that I was physically addicted. Not in four or five days, I thought. But life seemed to lose all its sparkle when I anticipated the rest of the day without smack. And the sparkle came back as soon as I imagined *one more time*. In fact the whole world seemed to light up. Maybe it's just about loneliness, I wondered. Jimmy and I are friends, and I don't know when I'll see him again. But the thought of another evening hanging around with Jimmy had no appeal without the added attraction of getting high.

The brain changes with addiction. Not in one or two systems, but in dozens. Neuroscientists are still trying to crack the problem, and each year they find more changes: changes in dopamine flow, changes in sensitivity to dopamine, changes in other neuromodulators such as acetylcholine, changes in the striatum, the amygdala, the hippocampus, the hypothalamus, and profound changes in the prefrontal cortex, the seat of appraisal, judgment, and consciousness itself. Changes in the brains of lab rats feasting on their daily supply of free narcotics. Changes in the brains of humans, imaged with fMRI or PET scans. Brains of heroin addicts, coke addicts, crack addicts, meth addicts. Who get thirty or forty dollars for their participation, a nice contribution to their day's income, to help pay for the habit that brought them there. Brains of recovered addicts that still show a sharp spike in activity when pictures of paraphernalia are flashed on a screen. Some of these changes begin much sooner than was previously believed. In days, as I was now discovering. Some are temporary, but some last for months, years, maybe a lifetime.

We shouldn't be surprised. Learning to play the piano or violin changes your brain permanently. For example, violin players show increased volume in the part of the motor cortex that controls the left hand. Driving a taxi changes your brain. A famous study using MRI imaging showed that the hippocampus of a typical London cab driver is quite a bit bigger than average. Memory, whether of city streets or positions on the neck of a violin, is the result of learning. And learning increases the number and strength of synapses connecting particular brain cells—those most activated by the incoming information. Yet in order to learn, you also need to forget, to make room, to stop seeing something one way so that you can start seeing it another way. Which means that other synapses— those least relevant to the new information—dwindle and even vanish. So learning strengthens some synapses and weakens others, and those synaptic changes remain in place long after the learning first occurs. Long-lasting synaptic alterations make the brain a habit-forming machine. That's the legacy of learning. If the brain did not form habits, if synapses didn't adjust to new events and *remain* adjusted, then we would have no memories, no skills, no biases, no knowledge.

Yet learning doesn't work alone. It has an ever-present sidekick as it goes about its job of synaptic sculpting, and that sidekick is emotion. Powerful emotional experiences change the brain rapidly and permanently. Fear and anxiety accelerate learning about things to avoid, a big part of the recipe for obedience in childhood, caution on the highway, and phobias about just about anything at any age. The emotion of desire facilitates learning about things to acquire. It holds onto those things with a kind of neurochemical adhesive, keeping them dead centre in the mind until they become automatic and habitual. Addiction takes desire to an extreme. In fact, an addict's desire for drugs and a starving animal's desire for food have a lot in common, neurally speaking. Studies of food-deprived mice reveal a barrage of chemicals cascading around the nervous system when the search for food becomes desperate. And foremost among those chemicals, not surprisingly, is dopamine. Good old dopamine, the chemical mover that gets us to chase after whatever it is we want, whatever spells relief. For starving animals, dopamine makes the brain a vehicle for seeking food; for addicts, it sends the brain hunting for drugs. In fact, dopamine-powered desperation can change the brain forever, because its message of *intense wanting* narrows the field of synaptic change, focusing it like a powerful microscope on one particular reward. Whether in the service of food or heroin, love or gambling, dopamine forms a rut, a line of footprints in the neural flesh. And those footprints harden and become indelible, beating an intractable path to a highly specialized—and limited—pot of gold.

If emotions join up with learning to modify the brain, then we shouldn't be surprised that a drug of abuse, a drug with enormous power to generate wellbeing when it is present and craving when it is absent, changes the brain profoundly. A drug that satisfies a kind of psychological hunger, temporarily, then leaves you starved for more of the same. A drug *designed* to change the brain, at least for a while, because after all that's what drugs are for.

As to where these changes take place, the bottom floor of the prefrontal lobe—the orbitofrontal cortex (OFC)—is a good place to look. In earlier chapters we saw that the OFC, pumped up with dopamine, assigns value to things: things like Lisa during my dalliance in New York, and things like heroin now. It anticipates how something will feel: whether it will be rewarding or upsetting, pleasant or boring, "good" or "bad." Orbitofrontal neurons tune themselves to resonate with emotional meaning, fertilized by dopamine pumped up from the VTA. And each time they are activated, that meaning emerges more vividly, because of the ongoing sculpting of synaptic connections—learning fuelled by desire.

With the emergence of addiction, the orbitofrontal cortex divides the world in half: good and not-so-good. The anticipation of the drug, the foreshadowing of that special feeling that comes each time, the synaptic imprint by which your brain recognizes *that* drug, *that* special feeling, is the purr of your OFC turning on and tuning in. Wow! Cool! This will be great! Those orbitofrontal neurons not only feed a message to "you"—the user—they also send excitation, in the form of glutamate, down the tubes to many other regions. In fact, those downstream effects are a big part of the feeling, the nuts and bolts of that emotional meaning radiating through brain and body. Amazingly, one of the early targets of this glutamate offensive is the dopamine factory itself, the VTA. This causes the release of more dopamine. In other words, dopamine flow to the orbitofrontal cortex elicits a wave of activation that ends up releasing *more* dopamine, cycling back to the orbitofrontal cortex: a classic vicious circle. Many brain processes turn out to be vicious circles or feedback loops—a design principle that explains a lot when it comes to addiction. But this particular loop includes an additional player: the ventral striatum, that hub of excited seeking. The glutamate tide coursing out from the OFC bathes the ventral striatum as well as the VTA. Now recall that the striatum feeds on dopamine as well. That's a big part of its diet. So the ventral striatum finds itself suspended between two colliding waves: glutamate from the OFC, carrying meaning, and dopamine from the VTA, carrying thrust. That's how the ventral striatum becomes the henchman of the OFC, ready to do its bidding, while both members of this little gang keep priming the dopamine pump that keeps them both humming.

Dopamine creates engagement with life's pleasures—both natural ones, like the taste of cheesecake, and unnatural ones, like the pulverizing fist of narcotic sedation. But when those pleasures are out of reach, when the goal is beyond your grasp, two things happen. First, if the goal remains attainable, anticipated but not yet present, dopamine flow gets stronger, energizing pursuit, tuning orbitostriatal connections in the moment and entrenching those same connections over minutes and hours. In this way, orbitofrontal *value* is translated into striatal *craving*, and, with repetition, the value–craving amalgam consolidates into a lasting union, a dependency that drives away the competition, perhaps forever. When the object is just out of reach, that gush of dopamine feels like raw desire, a deep itch, the contraction of an incomplete soul—whether the object of that desire is a girl or a drug. The second stage is when the goal is no longer anticipated, when you've given up. This stage brings the addict face to face with the world's other half: the not-so-good half. Because when drugs (or booze, sex, or gambling) are nowhere to be found, when the horizon is empty of their promise, the humming motor of the OFC sputters to a halt. Orbitofrontal cells go dormant and dopamine just stops. Like a religious fundamentalist, the addict's brain has only two stable states: rapture and disinterest. Addictive drugs convert the brain to recognize only one face of God, to thrill to only one suitor. And without that purveyor of goodness, orbitofrontal neurons become underactivated, sleepy, deadened. So the glutamate tap gets turned off. And, as a result, dopamine flow goes back, not just to a trickle but to less than a trickle, because the dopamine factory now relies on its supersized boost of glutamate, brought in fresh daily from the OFC, in order to maintain production. This is key. The net result of having an over-specialized OFC—one that is either enthralled or asleep—is that the ventral striatum follows suit, becoming underactive itself when the drugs have run out: because there's not enough dopamine to pursue goals, and not enough meaning to care. So the world of other things—of *everything else*—becomes dreary indeed. And for me, the image of a day without smack stretches out like an endless expanse of wallpaper. It's not Jimmy that I need, it's heroin. After only a few days, I'm nearly lost without it. A boat without a sail. A car without an engine. A hiker looking across a landscape with no hills, no valleys, no seas or rivers or lakes. Everything is flat. Until I hit the escape button and say "just one more time," and the orbitofrontal engines come to life again.

I tell myself that Jimmy is my friend. I love him and I love the roller coaster of wicked fun we're riding together, my buddy and I. But Jimmy is hardly a person to me. Jimmy is more like paraphernalia, a human face on the accoutrements of heroin. I'm probably the same thing to him. This realization lurks just out of reach, but its rumblings darken my thoughts and cloud the day still further. Hanging out with Jimmy without getting high, listening to his chatter, enduring his moods, would be . . . infinitely boring. So, each day, in the wan hours of the afternoon, Jimmy and I come to the same conclusion: let's get some more.

Scoring wasn't always easy. Sometimes we'd have to try three or four places. But we'd always succeed in the end. And the chase, the act of procuring, was exciting and tantalizing—thanks to the swirling crescendo of dopamine that foreshadowed its arrival.

A song popular on the radio at the time had these words: *Get down! Get down! Get down to the people, get down!*

The political implications were lost on us. Getting down was slang for getting loaded on smack. Smacked, smacked down, bashed flat into the floor with the sledgehammer of the world's most powerful opiate. We'd drive from one potential score to the next and drown out the radio with our gleeful, screeching voices: "GET DOWN! GET DOWN!"

Despite his assurances, I didn't actually believe that Jimmy would pay me back. He and I were bound together by something more primitive than friendship or trust. We were connected by a compulsion, one that gathered power day by day. We both understood at some level that this was our only real bond. Yet friendship was not entirely absent. We shared not only the heroin but also the vacation that came with it. We played in the warm eddies of our opiated coexistence. We shared stories. We watched movies. We cruised around San Francisco, picking up fresh shrimps at the Wharf or Chinese food in Chinatown. We went to the beach and gazed at the waves. We ate and slept together. We even thought we'd try having sex. Neither of us was technically gay or bi, but it didn't matter. Let's see what happens. Of course nothing happened, not only because we weren't drawn to each other, but because your libido sinks to zero on smack. We lay side by side in my father's big bed, slumbering through the night. If there was warmth there, beside the drug in our veins, it was hard to find.

When my money was really gone, I faced another moral cliff. It wouldn't be so difficult to borrow more if I set my mind to it, and the person who came to mind was Connie. Dad was out of town. I'd borrow a couple of hundred from her and get Dad to pay her back.

I told her it was for a badly needed winter coat. Dad would cover that. I knew it was wrong, but I simply didn't want to think about it. This was practicality. It would work. That's all that mattered.

Connie's money bought us three or four more days. And then it was over. There was no more money and no feasible way to get any. The depression hit before the smack was completely gone. There was one last night, a sort of goodbye party, but tomorrow looked impossibly bleak. The realization of just how low I'd sunk dawned slowly, hour by hour, like waking from a coma. The next day, Jimmy left. There was little in the way of goodbye. I sat in Dad's apartment stunned. I was absolutely broke. I could not afford groceries. I could not repay Connie, or Dad for that matter. A desperate remorse descended. I turned on the TV and started drinking.

By next afternoon, the withdrawal symptoms start to hit. Withdrawal symptoms! Are eight or nine days really enough time to get physically addicted? Oh shit. I guess I miscalculated. My skin is clammy with sweat, my eyes are tearing, and my nose is running. And running, and running. I have to wipe it constantly. It's getting chafed and raw. I've heard of this happening to other people, but now it's happening to me!

I pace around the living room, sit down, stand up, trying frantically to think, to reel in the spreading wrongness in my body. I feel hyper-aroused, way too alert. My breath comes fast and shallow and my heart races whenever I get up off a chair. Every cell is over-activated, set too high, buzzing, yearning, vibrating with the rhythm of my accelerated pulse. *Thump, thump, thump*, each heartbeat sends waves of energy down channels already shredding with strain. Limbs feel stretchy, tense, overextended, wanting to contract but unable to. I want to disappear. I don't want to have a body. This one is useless for anything but twisting discomfort.

Can't quite catch my breath. Don't have enough air to support the bellows of exertion, exertion leading nowhere but back to this reservoir of relentless stimulation. Can't think clearly. Can't find myself. No self here. Just yearning. No peace, no forgiveness, no compassion for the flopping, dying fish on the floor of this boat, caught by its own stupidity, unworthy of pity.

And then my stomach starts to churn. Cramps explode into diarrhea and I run for the toilet. And then again ten minutes later. And then again. What is this monstrous disease? What has caught my whole being in a relentless attack, a savage incursion into every system, mind and body? A wash of sensation whose whole point seems to be to make me suffer, make me pay?

The source of opiate withdrawal symptoms can be seen as an *anti-reward* system that finally gets a green light after waiting too long with its engine racing. The idea is that the body needs to maintain its equilibrium. When stress or pain push it in one direction, internal opioids (and other neurochemicals) bring it back to a zone of relative comfort. But when opioids are ingested in huge quantities, pushing the body far from equilibrium in the other direction, when the body is much too relaxed, too soothed, too sedated, the counterforce pulls in the opposite direction. And when opiate sedation lasts for days at a time, there is a buildup of this counterforce, a gathered momentum, whose purpose is to drag you back in the direction of stimulation, alertness, and tension. Because that's the direction in which equilibrium now lies. The only trouble is that you spring right past equilibrium—far beyond it, in the wrong direction.

Opiates ease stress and anxiety, and a primary mechanism for doing so is the active suppression of CRF, corticotropin-releasing factor. CRF is the chemical that shoots down from your hypothalamus to turn on your sympathetic nervous system: fight or flight. As part of that response, CRF stimulates adrenalin flow, a big player in the physical readiness that allows us to respond to challenge—with energy—to run away or turn and attack, as necessary. So, when the opiates run out, CRF rebounds in a massive surge, like a sleeping giant suddenly back on its feet. The result is an overabundance of energy. Way too much energy for my weary flesh. CRF rebound is the physical essence of anxiety. After all, anxiety, which is really fear spread out over time, turns on the sympathetic nervous system naturally, getting you ready to defend yourself or at least remain on guard. Its side effects include rapid pulse, shallow breathing, sweat, dilated pupils, and the razor-like sensation that things are very wrong. A major problem for opiate addicts is that the hyperactivation of CRF creates the very sensation that leads to recurrent use. A heightened CRF response means an exaggerated *feeling* of stress. And that's particularly dicey for me, like most addicts, whose lives would be stressful even without an overactive nervous system. The result is a blast of CRF as soon as the drugs wear off. Enough to drive even the most resolute individual (hardly my claim to fame) back into being an addict. Because that's the only way to bring CRF levels back to tolerable levels.

While the body is whipped around by the *physiology* of anxiety, withdrawal also magnifies the *psychology* of anxiety, just to complete the torment. Norepinephrine, the neuromodulator that turns on alertness in the brain, is part of the antireward armada, and it rebounds as forcefully as CRF when opioid suppression wears off. That's just way too much alertness for a consciousness that still wants to recede from the world, to find shelter from the world. Even a normal level of alertness would be uncomfortable right now—but this? Norepinephrine rebound has a nasty effect on the amygdala, the organ responsible for the emotional colour stamped onto things and events. When my amygdala, blissfully vacationing for the last eight days, becomes swamped with norepinephrine, the emotional colouring of things and events becomes greatly amplified, and most of that colouring is unmistakably dark. The world is flooded with frightening emotional memories and sensations, painting just about everything with dangerous hues—exploding darkness. Anxiety about being alone. Fear of being rejected, hated, abandoned. Flashing images of ugliness and self-harm. The vileness of self-injection. The worm of self-contempt growing, growing. Fear of the physical changes that, after all, probably won't kill me. Anxiety about having no job and no money. Anxiety about my father's outrage when he finds out how I lied to Connie. My amygdala, gorged on too much norepinephrine, joins my hypothalamus in a pincer movement of mind and body—yielding anxiety, stress, and physical discomfort—the fizzing negative afterimage of a sedation too long and too deep.

I tried to drown the symptoms in alcohol, which helped sedate me but did nothing for the depression or the physical symptoms. Dad was coming home in another day or two. I had to clean up the apartment and get rid of the evidence.

Jimmy called the next day. First he spewed a stream of ideas for getting more money, if only I could put up enough for another day. When that didn't work, he begged me. For twenty dollars. He said he was sick. My deadened emotions suddenly flared into anger.

"Fuck you, Jimmy. Don't talk to me. Don't call me. I'll give you ten bucks, which is all I have. I'll leave it under the mat. I don't want to see you again! Ever!" I slammed down the phone. I put the money into an envelope and left it for him. I had to walk. I was afraid that I would change my mind and call back. When I got out to the sidewalk, I crushed my syringe under my boot, tears streaming, then threw the pieces in the trash. A gloriously sunny San Francisco sky mocked me. I felt pathetic, writhing under its gaze. I'd walked half a block when a sudden wave of fatigue nearly dropped me. I crawled back to the apartment and slept.

Dad and Connie got married in late February. There was a huge party in their backyard, following a ceremony led by a robed practitioner of some exotic religion. I ended up staying over, picked up my stuff the following week, and then moved in officially. Connie's big, rambling house felt like a four-star hotel, perched halfway up one of San Francisco's highest hills. It was inhabited by Connie's three mixed-up children: Marcus, who was my age and a dedicated communist; Laura, a few years younger but avidly pursuing Jesus; and little Sarah, who would grow into an anorexic and then a coke addict before settling down. We eyed each other warily when we passed in the hall. But the house had many rooms. After a few nights on the sofa, fending off a foul-tempered terrier despised by all but Laura, I finally got a room of my own.

Edgewood Avenue was a red-brick street, lined on both sides with cherry trees that had already started to bloom. A pretty schoolteacher named Hannah lived in a cottage behind the house next door. We'd met at the wedding feast, where she warmly invited me to come for a visit—anytime. One evening a couple of weeks later, I got up the nerve to ring her bell. To my great astonishment and delight, I wasn't permitted to leave until the next afternoon. I started dropping in on Hannah every few nights. She would greet me at the door in a see-through nightgown, arms folded, smiling. She seemed to define some unique niche between bossy and lascivious. After a few conventional pleasantries, we'd go to bed, where she coached me, lesson by lesson, through her self-made curriculum of sex for beginners. I didn't mind repeating the basics until she was satisfied with my performance. I was growing to like it here on Edgewood.

But Dad found out about the money I'd borrowed from Connie, and he was indeed outraged. Before packing up Broderick Street, he tore the Indian Room to bits in a fit of anger. That was his way of Saying No to Drugs. And then he turned his wrath on me. It was an emotion I'd rarely felt from him. He yelled at me, called me names, and then held me by the shoulders and laid down the law.

"You will get up in the morning with the rest of the family. You will come home every day by six o'clock and have supper with the family. And you will go and search for a job, every day, until you find one, and spend your time productively, volunteering somewhere, until then. And there will be *no more hard drugs* while you live in this house."

I looked down and nodded.

He raised my head to look at him. "All that's left between us is love," he said.

He had never talked to me like this, with such authority, such conviction. I was shaken. I leaned my head on his shoulder and cried while he held me. But instead of sinking into depression, I felt myself buoyed up by an entirely unfamiliar sensation. I felt . . . safe. Cared for and safe.

For the first time since the age of fifteen, I believed my father loved me.

• PART THREE •

GOING PLACES

TRAVEL BROADENS THE MIND

E very few weeks, for most of a year, I'd entreated my mom to come for a ride on my motorcycle. One evening in June 1971, she finally consented. I roared up to the front door of the house she shared with Fred, as I had many times before. I felt welcome here— to a degree. It wasn't just Mom's place; it was Fred's also. Fred was a good man, but who wants a disenchanted twenty-year-old hanging around your love nest? And Mom and I had a complicated relation- ship of our own. I felt squirmy, sometimes panicky, under her barrage of questions, painfully direct pep talks, and exhortations to be a better human being, a better man. She wanted me to be comfort- able in my skin, and I just wasn't. She wanted me to be stronger than I was and more honest than I was. So I tried to make myself strong, to *seem* strong, which failed both standards at the same time. Being strong in my mother's eyes had been my goal since roughly the age of five, when our Kodak Brownie memorialized me, standing in a white bathing suit at the top of the jungle gym in our backyard, flexing my muscles, playing Tarzan. Even then I pretended to be stronger than I was. But now being strong meant staying off drugs. Was I that strong? Ironically, our motorcycle ride that evening would start me sliding down a more familiar path.

I was to leave for Malaysia in three weeks. Dad had gotten a fabulous job as a doctor for the aboriginal population who lived in the highlands of the interior. He and Connie were already there. They'd rented a house, a bungalow in a jungly suburb, and I was invited to come and stay with them for six months to a year. To help me get my head together. To complete my home schooling in responsibility, maturity, and decency. I couldn't imagine anything more exciting than a year in Asia, whatever the fine points of the contract, whatever the job specifications, the lifestyle. But that meant I had to say goodbye to California: goodbye to my motorcycle, goodbye to Mom and Fred, and goodbye to Abbie.

Soon after I dropped out of Hannah's tutelage, I fell in love with Abbie. Suddenly, seriously in love. I floated through several months of my life on Edgewood in a state of continuous adoration. Abbie seemed impossibly beautiful, loving, honest, and pure. She was slender, almost boyish, with firm breasts and dark blond hair, and she played the flute. Her lips seemed always ready to smile, opened slightly as if to play, and her cooing alto voice epitomized girlish grace in my eyes. And she *was* a girl. Not quite a woman at eighteen. A little damaged by events she would not or could not reveal. Which gave me the opportunity to rescue her, to be the strong, grown-up man who would save her from her demons. Yet I couldn't say no to Malaysia for her sake. And anyway she was planning to start university in New York in a few months. I'd go to Malaysia, but we'd reunite in a year. I'd come back tough, weathered, and wise, bronze-skinned and seasoned, with great adventures to tell. I'd be irresistible.

Once I got the Triumph started, Mom gingerly climbed on the back. Fred stood there smiling in the warm evening air, affable and teasing. And we were off. Down one street, up another, then up a good deal higher on the gorgeous curvy road that rimmed the tops of the Berkeley Hills and traced the canyons that cut through them.

Wind in our hair, my mom laughing and squealing with delight, the sky reddening across the Bay. It was great. Just what I'd wanted. Then, on the way back home, we entered an intersection without stop signs. We got there first, but the car coming from our right was going a lot faster. It hit us head on.

I woke up after a few seconds to the sound of my mother's screams. I was terrified. I crawled over to where she lay curled up. Her knee was fractured. Not too badly, as it turned out. I made soothing noises. She stopped yelling and pointed at my right foot, and that's when I realized that I should be yelling too. My foot hung from my leg at an unnatural angle. Then the pain started. Minutes passed in and out of awareness, and a lot of people gathered around us. I looked up at them through a haze of shock and pain. Fred was there, somehow, and he cut off my boot with garden shears as my foot and ankle began to swell. I hadn't realized it, but we were less than a block from the house. Then Mom and I were lying side by side in an ambulance, moaning, crying, and giggling a bit. It was just too weird.

After the surgery, I was told that my leg, ankle, and foot had been broken in quite a few places. The bones were now reconnected, some hardware had been inserted, and the next few weeks were to be spent healing. Just healing. I went back to Mom and Fred's with a cast up to my groin. I was going to stay there until my flight to Malaysia. I got the guest room, and both Mom and Fred catered to me, reminding me of childhood illnesses eased by tea and toast. Abbie dropped by every few days and buoyed me incalculably. Yet I was a guest in this house: that much was certain. I didn't live here. I had not lived under my mother's roof since the age of fifteen, and Fred left no doubt that this house belonged to him. So, while I still didn't question Mom's love, in the abstract, I thought I detected some distance from both of them. My motorcycle surprise hadn't turned out to be such fun after all. Mister Macho Guy.

The only thing I remember distinctly from that period is the Percodan. Those big yellow pills, two every four or five hours. They took care of the pain. They also gave me a glow of pleasure I had not felt since my adventures with Jimmy. By the second week of recovery, my need for the Percodan was wearing off, but my appetite for it was growing. Oxycodone, the main ingredient, is a potent "semisynthetic" opiate. A polished version of a heroin high—milder, to be sure, but also smoother at the edges, thanks to the wonders of modern chemistry. I began to save pills throughout the day so that I could have something of a rush when I took them in the late afternoon or early evening and then enjoy a well-orchestrated fadeout into dreamland at ten o'clock. My orbitofrontal cortex was reunited with its old lover, or perhaps a sleeker, younger version, with less clout but more class. My striatum warmed with anticipation as the afternoon drew to its close. I knew I was slipping. I think my mom knew it too. But I felt I was entitled. The pain was my ticket to get high each and every day.

I told myself that I deserved this after six months of being a perfect angel. What I didn't tell myself was that I relished the subterfuge as much as the drug. By sneaking a nice big helping of my opiate prize, right under the noses of my caretakers, I was declaring that I didn't need a great warm welcome. I didn't need to be constantly cherished. I didn't need access to the inner sanctum, to feel a deep sense of inclusion. I had my own supplies of warmth.

The Malays are small, slender, and brown, with delicate features. Beautiful soft eyes often crinkled in smiles of welcome or bemusement. What they saw when I got off the plane was a relatively large person, with white skin, a hippie-style Afro and beard, dressed in bulky white overalls, pulling myself along on a pair of crutches. There was a lot of gazing and grinning as I hobbled through the airport. Then there was Dad, looking bemused himself. We did the brief manly hug, he picked up my duffel bag, and we walked toward the exit. A wall of warm soup was waiting outside the door. And then there was the car, and we were off.

The house was on the outskirts of Kuala Lumpur, the capital of Malaysia. KL, as everyone called it, was a large city with a few elegant skyscrapers poking above streets stuffed with small shops and stalls. But on the outskirts the houses were expansive bungalows with white walls and red tile roofs, separated by lush lawns and clusters of flowering bushes the likes of which I'd never seen. Against the pale sky was a canopy of treetops, spreading out symmetrically, like cool-green fireworks, high above a layer of smaller trees, intricately limbed and adorned with blossoms. The air itself was perfumed. We drove slowly up the small street and into the driveway of a stately bungalow that looked no different from its neighbours, though each lay half-hidden until you drove right up to it. I stood on the large front patio next to Dad. Beside us stretched a field of banana trees.

"That's our orchard," Dad declared with pride.

I followed him inside and hopped along, mesmerized by his tour-guide monologue, as he led me from room to room. Connie was at the market with Ah-Kin, our *amah*. "An amah is both a cook and a housekeeper," he told me. "Yes, we have a cook, and she needs to help Connie."

"Help her?"

"Help her shop. Help her learn where to find everything . . . and how to get the price down. When they see a white face, the price doubles. More than doubles."

"Well, then, why doesn't Ah-Kin just go by herself?"

"Because Connie wants to learn. And Ah-Kin doesn't know what we like." Then, a moment later: "And we have the money." There was that pride again. Dad was enjoying his new role.

The family's bedrooms branched off the central living and dining rooms. Dad and Connie's room at the front of the house, Laura's room next, my room next, and then Sarah's room last, near the kitchen. Laura was sixteen and Sarah eight at the time. Ah-Kin's room was at the back of the house, past the kitchen, a region that seemed vaguely out of bounds. I was surprised at the elegance of the place. The floor was cool, polished stone under my bare feet (foot, actually, for a few more weeks) and the walls were a stately white. An archway separated the living and dining rooms. The furniture was wood or bamboo, with colourful cushions adorning graceful frames. Everything was new and everything was clean. Ceiling fans whirred softly above us. I felt a low-grade excitement, anticipation of an utterly different life. There was nothing familiar here. Except a slight awkwardness between Dad and me. I couldn't tell which of us it came from, and whether it was real or else a ripple in the wet air. When there was nothing left to say, Dad suggested I go and relax after such a long flight. I lay on the batik bedspread and looked into the housing at the centre of the ceiling fan, an eye in a tiny hurricane.

It didn't take long to adapt to the household routines. I had no responsibilities. Just books to read and a guitar to play, sitting on my bed at night with the door closed, careful not to disturb the others. Mealtimes brought us together in the dining room. Ah-Kin was a fabulous cook, and I loved the aromas emanating from the trays she brought to the table before she disappeared again. But dining with the family was at least as much of a strain as it had been in San Francisco. Maybe more, since there were no obvious escape routes. Nobody was very interested in what anyone else had to say. Sarah was cute and cuddly. She reported the latest news of her rabbit's escape from its cage, Ah-Kin and her in pursuit. Laura was a sullen teenager committed to Christian fundamentalism. Her silence was austere, sometimes judgmental. Connie was always chipper. She told us what she'd discovered in KL that day. New shops or stalls, Chinese markets, bickering amahs, barrels of indefinable pickled things. Crowds everywhere. She laughed at her own descriptions. Connie was nice to me. She was trying. But when I wanted to share anything personal, it didn't translate. Nothing got across. Something seemed to be missing in Connie. Maybe she was just tired of parenting. Yet parenting wasn't what I wanted. I wanted someone to talk with.

And then Dad would come home in the late afternoon, shower and change his clothes, then stand there in pristine white shorts and T-shirt, signalling the arrival of cocktail hour. He and Connie gathered the ingredients: gin and tonic, salted crunchy nuts, and crackers with soy flavourings. I'd stand around, hanging from my crutches, feeling useless but salivating now in anticipation of that first sizzle of gin. We'd take our seats on the patio and gaze across lawns and bushes at the pink undersides of clouds, benign once more after the daily downpour. We were following in the footsteps of our colonial forebears, Dad joked. Dinner would come along when we were ready.

He told us all the interesting things he had seen or done that day. At the beginning of each new episode he looked around to make sure we were listening. I cocked my head at an attentive angle. I asked thoughtful questions. I wanted to provide a receptive audience. Connie simply gazed at him, nodding and laughing on cue. I couldn't figure out what to say for my turn. In fact I couldn't figure out when it was my turn or how long my turn was supposed to last. I wasn't doing much of anything these days. The really big events in my life were either fifteen thousand kilometres away, in the person of Abbie, in the future, waiting to be revealed to me, or else hidden away in my fantasies, most of which I wasn't keen on admitting to myself, let alone sharing. So I rhapsodized a bit. I was the poet-scholar brimming with exalted ideas, or the wandering minstrel, stopping to rest while my wounds healed, dragging my guitar around. I was the philosopher, the thinker, the dreamer. I had been reading a New Agey self-help guru, Krishnamurti, who extolled self-awareness and advocated a detailed analysis of one's own thoughts and motives. Self-analysis came naturally to me. I could see its virtues, but it scored pretty low in audience appeal. And I guess, because of the inadequacy of the story line, Dad's impatience grew palpable before I'd gotten very far. He had stories of his own left to tell.

To bolster my confidence, I would refill my glass whenever I got the chance, when it didn't seem too obvious. I would stop by the fridge on my way back from somewhere and pour more gin into my glass, the tonic following quickly to hide its level. I would come back out on the patio and sit, my leg up on something. Was it okay? Was I okay? Did Dad notice my speech thickening? Our eyes never met for long. Did that mean something? Back in my bedroom, I tried to find some pattern in the crazy rhythm of the crickets outside my window. I read, or wrote, listened to music and made up new songs, or drew pen-and-ink figures, those precise lines captivating me with their simple definition.

Every few days, I went with Dad to the hospital. It was hard to get around with my cast and crutches, but the place was endlessly fascinating. Gombak Hospital only treated aborigines—the *Orang Asli*, or "original men"—a race of small wiry people with fine, sculpted features, curly hair, superb musculature, and faces open wide to the world. The patients came in bunches. It would be cruel, unthinkable, to send your loved one to the capital, by jeep over rough roads or by helicopter, far more terrifying, especially when ill, without the protection and embrace of family. The Orang Asli spoke little or no Malay, and certainly no English. They depended on each other and on the nurses and orderlies who had learned their unique tongue. We depended on interpreters to understand the first thing about them. I couldn't imagine that the doctor-patient dialogues were particularly accurate. But everyone worked together, everyone tried, and that gave the place a warm glow.

The wards were brightly coloured dorm-like buildings, lined up in a ragtag row across a compound from the main hospital building, where medicines were kept and surgeries performed. The place looked like a summer camp. To accommodate the Orang Asli's advanced sense of caring, the interiors of the wards were always up for renewal. Beds were pulled together in new configurations, three or four abreast, forming little encampments topped with a platform of used-looking sheets. On these platforms lay the patient, flanked by the spouse, the mother or father, maybe an aunt, a sister or a brother for safe-keeping. It would be unheard of to isolate the sick one, whether hacking with tuberculosis or delirious with malaria. Germs were not the issue. Spaces between the clusters demarcated families, and everything got reshuffled each time a new family arrived. While the patient sat or slept, family members went back and forth to the massive fire pit, where they cooked their own food and socialized. No mashed potatoes and peas for these folks. They brought back sizzling skewers of meat and other delicacies, which probably helped cure the patients as much as any treatment. Or gave them reason to live. The wards were hives of chattering children playing amid household belongings, fabrics, artifacts. Unlike the straight lines and silent, featureless corridors of our hospitals, they were brimming with colour and warmth. And above all that noise you could hear the sweet trill of a nose flute, as some old man, his face a fugue of wrinkles, sat and played the instrument he had lovingly carved an hour before.

I limped around the hospital, looking, learning, chatting, while Dad did his doctoring. I got to know some of the interns, the Peace Corps guys, the Malay nurses, and I became a familiar sight to them. The Peace Corps doctors lived in huts across the river, and they crossed that river, commuting to work, in about three minutes, by way of a huge tree that had fallen with a little help exactly where a bridge was needed. They teased me: could I walk across the bridge with one and a half legs? Not on my feet, I couldn't, but I straddled that log and crossed it on my bum, earning some mixture of amusement and respect.

Back home, after everyone had gone to bed, I'd try to compose another letter to Abbie. I had no trouble spinning out fantastic descriptions of the hospital and the Orang Asli. But it was hard to say what I was doing here, what I was feeling. We're a family of strangers, I'd begin, trying to muster some existential gravity in place of the adolescent angst I actually felt. But no matter how I pitched it, my life didn't feel much like my life. Why were they taking care of me? Didn't I owe them something in return? It was only six or seven months since the heroin debacle and the *big lie*. But I felt more guilt and shame now than I had all those months living on Edgewood Avenue. And that's because I was starting up again. There was no denying it. I had pilfered a few codeine tablets from Dad's medicine cabinet. How many? How often? I wasn't really keeping track. I could pretend they helped with the dull ache I still felt in my leg, but I knew that pain wasn't the issue. I could feel the engines coming to life, more and more every week. Partly resulting from the reset that had clicked in with Percodan, partly in answer to the unfathomable dynamics of my father's house, I was getting ready.

So far, my excesses were mostly at the level of fantasy. But the more I fantasized about drugs, the more I felt I had to hide; and the more I had to hide, the harder it was to connect with Dad or Connie. Shame and alienation came hand in hand, and all I could do was try to keep a lid on it, by reminding myself, day after day, of my promise to "be good." I had vowed to various people that I'd go on being good for a very long time. I'd be good for Mom, and she'd see that as an achievement, a demonstration of strength. And for Dad. He'd see it as a sign of my independence and maturity. Yet, when I thought about it at night, "being good" for my parents lacked any real satisfaction. There was no pride to be had, no sense of accomplishment, because it wasn't for me. There was nothing in it for me. It just served to underscore my lack of independence. I hung on as best I could. I listened to Dad's stories. Raptly, enthusiastically. I asked Connie about her day, about her life. And I said less and less about my own untidy thoughts.

Those thoughts popped up like mushrooms on the lawn. Little abscesses on the skin of my good-boy persona. Threads of half-formed plans. Images of depravity. It annoyed me at first. Here I was, being as good as gold. I knew there were drugs all over the place. After all, Gombak was a hospital. And KL was full of seedy-looking drugstores. But I could hold on, keep low and cool. Until I woke up from a dream about heroin: I'd been scurrying around, looking for a place to shoot up before . . . before something stopped me. Just get it into a vein. Get it inside me. Quick, before waking. And those dreams would inevitably start the wheels turning, stocking me up for a day's worth of rumination. That just wasn't fair. A rebellious voice began to mutter that the whole notion of recovery was a scam. Once you recover, aren't you free to move on? Isn't that the whole point of *being free,* a state I had tried to cultivate in various forms for as long as I could remember? Besides, what I took into my body was my own business, wasn't it? I had been so good for so long that I was sure I knew how to do it, to keep it under control. That was the main thing: keeping it under control. The occasional slip wouldn't hurt as long as I kept it under control.

Toward the end of my second month in KL, three things happened in succession. First, my lifeline of letters from Abbie dried up and stopped. Her letters had been rich with declarations of love. And they'd kept the channel open for my own restless mutterings. Who would I talk to now? I had a drawer-full of notebooks with elegant batik patterns on cardboard covers, and I turned there to record my moods and the maze of thoughts that came with them. I analyzed everything. I had kept Abbie in a secret room in my mind, a little furnace room radiating warmth out to every nerve ending. But security was a trap, wasn't it? According to Krishnamurti, security was a prison that held you in the past, encased you in the known, barred you from the ineffable currents of the present tense, the creative emptiness of now. I analyzed my need to prove something to my parents. That was easiest with Mom, perhaps because she was so far away that there was no day-to-day data to mess up my theorizing. And I analyzed my need for Abbie. Maybe I was better off without the letter writing, the constant immersion in a dreamy narrative. Maybe it was okay that she was seeing other guys. Just till I got back. We both had some exploring to do.

Second, we removed the mini-cast I'd been wearing since the big one came off weeks before. I could walk freely now. But my leg remained weak and my muscles slack and lazy. I stayed home most weekdays, walking back and forth from room to room. And those were the times the house was most often empty. Temptation city! Dad's medicine cabinet called to me relentlessly. It was always there, ever-present, waiting, whispering. So I made little forays. Four or five codeines, good for a few hours. He seemed to have a vast supply and wouldn't notice a few missing now and then. Codeine gave me a buzz, a break from the uncompromising shades of the day. But it was nothing compared to the silky glove of Percodan. So I kept looking, and sure enough, at the back of the shelf, I found something I hadn't seen before: a bottle of methadone tablets. This prestigious member of the narcotics nobility wasn't all that common in those days; I had no idea why Dad or Connie needed it. For pain? To ease the with-drawal symptoms of some friend of the family who wasn't around? Methadone is slow to come on and slow to dissipate, leaving many hours of dreamy sedation in its wake. I tried it one day, and then again a week later.

Then the third thing happened: a final letter from Abbie, to say goodbye. Despite Krishnamurti's advice to dismantle the habitual quest for security, I just wasn't ready.

"I love Eric very much." The words leapt out at me from the middle of the page.

Eric?! Who the fuck is Eric?

I threw the letter down and bolted out of my chair as though slapped. My practised detachment wasn't up to this. But I was deter-mined, almost as a punishment, to face it head on. I forced myself to acknowledge, again and again, that I had just lost the only girl I'd ever really loved. My glorious sojourn in Asia had ended up like an adolescent pop song. And out of my own stupidity! You don't just take a year off and assume that your beautiful girlfriend is going to hang around waiting for you. I wrote compulsively, trying to catch up to the present moment. Even if you flew to New York tomorrow, I told myself, you couldn't get her back. I cried. I wrote more in my journals. I tried to return to Krishnamurti's wisdom. I was better off without packing a security blanket, wasn't I? Free to roam the world with no return ticket? I'd thrown off my physical crutches; now I was throwing off my emotional supports. I was a budding Buddhist, a seeker, suddenly released from the bonds of attachment. But the next moment I was a heartbroken kid, crying in the dark.

Now I was really on my own. When the girls got back from school I'd close the door to my room. I would go back to my desk to re-read Abbie's last letter, searching for missed crumbs of affection, hidden insinuations that she might still be waiting for me. But I had missed nothing. I would begin my pacing again, now in a cloud of drug-soaked images. They infiltrated my reveries, these birds, veering away from the flock, spreading and forming into shapes of their own design. I wanted to take something powerful and wicked, to move far beyond the borders of my cloistered existence. I wanted to put something into a needle and feel it take hold of me, pick me up like Dorothy's tornado and drop me somewhere altogether different. My internal dialogues became increasingly harsh and vindictive. You're obsessed. You know that, don't you? Just do it, shithead. Instead of worrying about it all day long: Should I? Shouldn't I? Should I? Shouldn't I? That's stupid. Just do it. Pierce the wound, let it drain.

Finally, after a restless night, I found myself at Gombak, ready to hunt. I had not spoken two words with Dad on the long car ride over. I was in my own world. There was no one to stop me, no reason to stop myself. I said goodbye to him and walked to the main building as he left for the wards. I was on the lookout, my eyes and soul narrowed, my steps quickened, dopamine-powered. Excitement poured into the back of my throat thanks to gouts of CRF, activating my sympathetic nervous system as the race began. I observed myself with bemusement. How can you be so excited about a complete unknown? A wild card? There may be nothing here but Aspirin. But even in this condition I found myself better company than I had been for days, maybe weeks. There was something almost noble, self-possessed and jaunty, in this little escapade. I wasn't trying to be good anymore.

Thoughts racing, strategies forming and dissipating, I walked through the main entrance, the world suddenly shifting from sunshine to gloom, then down the central hallway, past the administration offices, a look of official purpose plastered on my face. Where did they keep the drugs? Past the busy administrative area, I began to try the doors, one by one. I found empty rooms full of wooden cabinets, squat and dowdy leftovers. From colonial times, by the look of it. At each doorway I looked behind me, saw no one, entered, then closed the door. I went through drawer after drawer, methodically, checking each off my mental list when I found nothing of interest, nothing but medical paraphernalia, tubes, clamps, braces, bandages, shunts, and needles. Needles? I grabbed a few and stuffed them in a satchel. Now, if only I could find something to put in them. They were large, ancient-looking things—glass cylinders with a barbaric-looking plunger poised between steel finger-holds. They looked like props from a fifties horror movie—*Dr. Jekyll and Mr. Hyde*. I played it up in my head, dramatized the whole adventure. And it *was* an adventure. I was through being good.

Orderlies and nurses passed me in the hall, but no one doubted me. There was no hint of suspicion. Only polite nods and smiles. I must be a doctor or an advisor or an intern. I must have business here. How convenient to have white skin. And although I'd anticipated as much, it still surprised me. My cloak of immunity stretched further with each step, and my search grew bolder. I no longer cared if I was seen entering that imagined haven.

Here was a room that looked promising: empty except for several large trays, glassware glinting in the shadows. Drugs? I closed the door and looked around, emitting a little moan of relief mixed with glee. On a table lay several large trays crisscrossed with dividers, separating vials by name or category. There were so many. My heart raced. To be caught with my hand in the cookie jar might still prove awkward, especially if word got to Dr. Lewis senior. Did you order this, sir? Quite unusual, asking your son to pick it up . . . The rows were divided into cardboard compartments, each containing a single vial of injectable fluid. I searched methodically, row by row. I thought I would find morphine, or Demerol, or some other opiate. I really wanted that. My frustration and disappointment grew with each false lead. Tranquilizers, antibiotics, antihistamines, anticoagulants, this and that, not of interest. And the clock was ticking. If someone were curious . . . Until, almost at the last row—what's this? Methedrine. What's Methedrine?

It took a moment to recall. Oh yeah, Methedrine, a brand of methamphetamine. Speed. Pure pharmaceutical speed. I had tried meth once. Over a year ago I had snorted a little pile of white powder, and I'd liked it a lot. I had flown above everything for twelve hours. But then came the jagged crash at the end, and so the shining vitality of a meth high wouldn't be much fun without a soft cushion to land on, a tranquilizer, ideally a narcotic, like . . . The sound of the word "meth" brought to mind Dad's methadone—bland-looking white pills. Maybe a meth sandwich, Methedrine on the way up, methadone on the way down . . . maybe that would be okay. Maybe it would be really great.

I was brimming with crazy anticipation. But then there was that cliff edge nearby. Fear. Throbbing recklessness, misguided, urge to . . . what? Not here! I wasn't going to shoot speed in a storage room. I had no idea what injecting one of these vials might do to me. I'd bring it home and take it tonight, after everyone else was asleep. I opened my satchel and tenderly placed inside it four vials and a couple of little steel blades for sawing off the glass nipples that sealed them. A couple of bottles of distilled water, alcohol, swabs. I closed it with something like pride. There was an indefinable heroism in this act, in the wrongness of it. The unmistakable wrongness of stealing drugs from Gombak Hospital, an act of pathetic delinquency, was a ticket to freedom.

By that evening the process was well underway. Sitting at the dinner table, going through the motions, the sense of justified rebellion condensed a little more. It extracted fuel from the evening's show-and-tell, something that had seemed so important, an event for me to join—now repellant. Dad's self-satisfied sniff, his sidelong glance at Connie when Laura took the bait and got pontifical. Connie's superficial praise, aimed at whatever was being said, whatever came along, masking a profound disinterest. I will leave you behind, my lovelies, to attend to your charade. And venture off into my own world.

By nine o'clock, everyone has more or less disappeared into their lairs. By ten I say goodnight and shut my door. I haven't shot drugs in a long time, and the anticipation is a fist grabbing me by the collar and pulling me forward through time. I set down my paraphernalia on the edge of the desk and begin my preparations. Compared with the ritualistic steps you have to take with heroin, there's little to do. I clean the syringe in sterile water and wipe the steel barb with an alcohol swab. I load it with two vials. I have no idea of the proper dose. Why did they even use methamphetamine clinically? I guess to shock the heart back into action when it stops its pumping. I've never shot speed and I know it's dangerous. Too much at once could cause . . . what? A heart attack? A stroke? Fear surges upward, shearing through a cloud of excitement. Fear and excitement make natural partners. Both ride the same flow of neuromodulators, especially norepinephrine, painting every thought and deed with bright, brittle hues. Whatever the mix, I note its effect in my dry mouth and slightly trembling fingers.

I sit in a chair beside my bed, with an almost brutal determination to keep going, despite my reservations, my fears. I tie the belt of my bathrobe around my upper arm and expose the vein below it, squeezing my fist, watching the skin grow taut like a fledgling erection. I push the grisly-looking needle into the bulge of the blue vein and encounter a surprising resistance. This needle is dull! And it's huge. I didn't realize before what a large hunk of metal is attached to this arcane-looking device. Okay, so it's dull. So push harder. I do, and finally the skin breaks and the point drives down into soft flesh.

Trying to penetrate the second skin, the wall of the vein itself, I almost cry out with pain. I am revolted, but also determined. A couple of stab-like thrusts, and finally the vein is pierced. I pull back slightly on the plunger and a tiny geyser of blood follows it into the syringe. I'm home. I push the plunger in slowly. About half the liquid is gone by the time I feel the first thrum of intense excitement ripple through my gut. I could stop now, but I don't want to miss any of it, don't want to risk something less than overwhelming. So I keep the plunger going while my belly fills with light. I sit back, close my eyes, and explode.

The inside of my body fills with bright, white pleasure. Hyperarousal in my toes, coursing up my legs, radiating through my genitals, wrenching open my gut. But there is no fear anymore. Only delight. My stomach fills with energy, intense but sublime. I am coming apart. The darkness behind my eyelids fills with pulsatile red light. And now my eyes spring open, unable to contain it. I send my gaze across the room, where each surface, each corner of every article of furniture, thrusts itself into reality with unnatural presence, with pure effrontery, proclaiming its special existence, its absolute uniqueness. Every edge, every surface, every colour shines with its own essence. The world has become unnaturally clear. But also quiet, still, shimmering with significance, supremely present. All my senses are on high alert. A hyperreality permeates the physical space around me. Whereas heroin seemed to take me deep within myself, methamphetamine jacks up my centrifugal momentum so that I am thrust outward, to an enticing and compelling world *out there*. I feel like pitching forward right off the chair, to do . . . something. But there is nothing to do. Nothing except to revel in this blissful tide. I roll over onto the bed and stare inward and outward at the same time.

This excitement and potency, this sense of efficacy and joyful enthusiasm, result from large amounts of dopamine, pooling in the synapses of my ventral striatum. We've seen how dopamine fuels the primary function of the striatum: to pursue goals. It does this by jacking up the salience of rewards, highlighting their specialness, their appeal, so that they fully occupy attention, so that they shine, while translating that beam of light to a motive, a thrust—an urge to act, to pursue, to achieve. Amphetamine releases dopamine through several molecular processes in different parts of the VTA-striatum loop. For one, it releases dopamine from its storage sites by tricking a clean-up molecule to work in reverse, emptying dopamine back into the synapses, rather than back into storage, after it has done its work. For another, it turns off the self-imposed braking mechanism that inhibits the release of dopamine when it first gets launched from cells in the VTA. Amphetamines make it harder for dopamine to be reabsorbed. It just sits there, in the synapse, activating, activating, activating, as long as it possibly can. And dopamine enhancement is not the only effect of amphetamines. Norepinephrine and serotonin levels get jacked up as well, because they too are left to dawdle in the synapses without being reabsorbed. That's why methamphetamine— the strongest of all stimulant drugs, the king of amphetamines—gives you pleasure and not just unchecked attraction. The neuromodulator cocktail served up by meth includes everything we need to feel engaged, alert, stimulated, and relaxed at the same time.

But the subterfuge of drugs like methamphetamine culminates in one brilliant trick: the message of meth is that goals don't matter anymore. When engorged on dopamine, the ventral striatum acts *as if* it's pursuing goals and gloriously, magnificently, achieving them. Yet there are no goals. The excitement is bogus. It's free. It's detached from any actual internal desires or external accomplishments. The dopamine deluge is continuous rather than event-specific, and so are the secondary tides of norepinephrine and serotonin. These chemicals are normally paid out in small quantities, based on a strict exchange rate, delivered in momentary bursts intertwined with elements of perception and action. But now they disgorge in one unbroken blast. Yet I am not lost within myself, as I would be on opiates. Rather, this joyful exuberance is connected to the world. First to the shiny surfaces of objects in my room, and then to the inviting darkness of the night outside my window. I want to race out the front door and become a part of everything, to partake in the glorious energy collecting in the night, in this magical country dripping with life.

There is light at the bottom of the door to Dad and Connie's room, but I am past it in a moment. Out into the night. And here the scintillating hyperrealism is softened by liquid darkness. I am pulsing with energy. I am vapour. I can move without effort. I can inhabit any object I see. Like the damply breathing flowers hanging in great bunches from their branches. Like the crystalline condensation on the threading of the badminton net. I am intensely enthusiastic about all of it. I am awash with dopamine and its gift of *connection*. I wish I could tell someone how wonderful it is.

The night has no end. I wander from one dark street to the next on legs that never tire. I investigate alleyways and gardens that have called out to me before but which I've never dared enter. I reach the main road, turn right toward the city, and propel myself along the wide shoulder, where oil lamps throw sudden light on the faces of men and women squatting in the darkness, selling food wrapped in banana leaves. They appear out of nowhere and then vanish as I move past. For a moment I stare at them, awestruck, and they stare back, no doubt astounded by this white face suddenly looming out of the night. No one stops me, no one questions me. Other voices on the street come from phantoms.

My goals are epic. My plans are the stuff of a mythical quest. The night reveals itself to me, only to me. I have a deep and dominant understanding of what is real and what is important, and this is a radical departure from my weeks of apathy. Confusion is banished. Thanks to dopamine, everything is clear. Everything is included in my focal attention. There is no periphery, no background. There is only the continually bursting horizon of absolute relevance.

I return home in the early hours of the morning. I know that I won't sleep at all, so I set myself to ink drawings. These occupy me for hours, without boredom or fatigue. But then dawn starts to lighten the edges of my blinds. How will I pull off this next part? I have to somehow pretend that I'm just an ordinary person, a sleeper, waking up normally with the rest of the family when it's time to begin the day. I wait until an hour before my usual wake-up time of seven-thirty, then turn off the lights and get under my sheets. I lie there in the twilight of my room, unnaturally still, as if in suspended animation. I close my eyes. I feel a sense of triumph: I can do this. No one will ever know. When Dad knocks and pokes his head through the door, I groan and stir, caught in the theatrics of the moment. An hour later I appear for breakfast, but I can't eat, and I don't want to sit at the table with anyone else. So I take some toast out to the patio and sit quietly by myself. Later I throw it away. My eyes are hollow lanterns glowing with unnatural light. No one must see me like this.

The house clears the last of its inhabitants and the bustle finally fades to nothing. I am luxuriously alone once again. But the purity of my state is now compromised by tendrils of fatigue. I maintain full-out alertness, but with muscles now stiff with overuse. Shards of bright energy continue to arc, but shadows gather between them. So I set myself to the next task of this forty-eight-hour day. I reach behind bottles and extract the methadone from Dad's cupboard. Still there, thank God. Still lots. I count out four pills, ten milligrams each. That should be about right.

And it works, as well as I imagined. Better. The soft weight of narcotic sedation creeps over my jagged nervous system like a sunset. I can feel its soothing warmth, softening, melting, merging the pools of discomfort that have begun to accumulate. Soon the air itself has darkened and the gathering shards of self-doubt condense into a chorus of whispers, increasingly cartoonish and benign. A rollicking Disneyesque recital, straight from Fantasyland. I close my eyes, and immediately I am dreaming. Awake but dreaming. The imagery dances and cascades. Here in this waking dream, in a darkness still lit by streaks of unnatural lustre, I arrange and rearrange the patterns and shapes of my private kingdom, unpredictable, magical, potent but serene.

Over the next month I experimented with other helpings from the Gombak medication room. But nothing was as interesting as methamphetamine. Two or three weeks later I did another late-night meth extravaganza: two full vials instead of one and a half. The rocket ride out was even more dazzling than before. Storms of hedonic energy raged up and down my body. I had judged the quantity just perfectly, I thought. Until the headache started up. It came on fast and got blindingly painful within moments. I was terrified. The headache overshadowed everything else, and I thought I had done something irreversible. Is it a stroke? For a few moments I just hung on. Then the headache started to subside, and I began to imagine that I was still all right.

Or had I already blown half my brain and I was just too loaded to know it? And if I had, didn't I deserve it, after all? For stealing drugs from the hospital, for the hiding and the lying that I'd already taken for granted and that would only get worse, each deception building on the last? A dark cloud of self-contempt gathered quickly that night. Beyond the bad publicity, would my death matter to anyone?

But then came the reprieve: a wave of sheer rebelliousness carried me back to some kind of resolve. I had perfected my aloneness, and it conferred a sense of purity, a protective shield. Dad had his own life. He was the Big Doctor, saving the Natives. Abbie had her man to cuddle up to at night. My path cut away from theirs. It was up to me, I told myself with comic-book bravado, to explore the reaches of darkness. And I really did see myself as a kind of antihero. Oh no, your words can't stop me, I fired back at my hidden detractors. I don't care what you think. I've got what's mine. Your words don't touch me. This ride will be sacrosanct. I won't look back.

But I could not escape the shadows of guilt and shame for very long. The voices that I imagined scolding me, haranguing me, came from inside after all. They fed off of my own angry shock at the lengths I was apparently willing to travel in order to feel potent and strong, free and resplendent. I continued to rebuild my own prison, taking chances that showed a terrible disregard for my safety, my life, leading back to a small, dark pool of anger and despair, waiting for another chance. Drugs worked, so there was no way I could resist them. But they did not work well: they inevitably led to a more profound sense of loss.

That particular binge ended, the following night, with a determination to shoot opiates. All this messing around had left me with a hunger for heroin, or something very much like it. But all I had available were my father's codeine pills, so I crushed them up, poured the brownish powder into a large spoon, added sterile water, and boiled the mixture over a candle flame. I must have missed a vein, created an abscess or a bruise, and then injected more liquid into it, because the blood on my arm streaked across a darkening purple bulge and dripped onto the bathroom floor. I was saturated with revulsion and starting to panic. Someone would find the stains. I would wake up in a pool of congealed blood. Or I wouldn't wake up at all. No, the ride at that point was anything but sacrosanct. Despite my best intentions, I could not escape the sick smell of weakness and fear. I had not outpaced it after all. Not since Tabor, not since San Francisco. This whole twisted scenario left me with the unmistakable verdict of defeat.

CONSCIOUSNESS LOST AND FOUND

The following six months were a grab bag of adventures, haphazardly arranged along two themes: my wish to explore the outer world, like any other traveller, and my wish to explore the inner world, like nobody else I knew. I was infatuated with Malaysia. There was no end of novelty in KL. I walked all over the outdoor market at night, tasting unnamable delicacies, escaping from Chinese rock bands on outdoor stages, sitting at tiny restaurants, writing in my journal. And I was infatuated with drugs. Their scent was everywhere, in my father's medical connections and in the half-hearted legal restrictions of the society at large. I kept my eyes and ears open. I found something called chlorodyne, a foul-tasting liquid that was loaded with opioids—"for diarrhea," said the pharmacist with a smile. Various codeine mixtures, other closely related opiates, even pharmaceutical cocaine. There was a lot to explore for an intrepid druggie.

I finally got involved in helping out with medical research, as my father had wished. One of my jobs was to inoculate Malay and Indian workers with a new kind of antimalarial vaccine, shot through the skin on a blast of air that avoided the risks of infection and the need for sterilization. I gave people the shots and somebody else recorded their infection rates. For a few hours a week, I felt useful. I also began to study Indian music. I bought a sitar, found a teacher, and travelled to a city eighty kilometres from KL every two weeks for my lesson. My teacher was a fortyish Indian woman who had studied with some low-ranking musician and learned the basics. She was a middle-class housewife, showing the proper protuberance of flesh in the gaps of her sari—no virtuoso, but I wasn't a demanding student, and I hungrily memorized the Hindi characters that stood for the notes of the many different scales. And I went on jungle trips. These were the most memorable of times. They were physically demanding, a little dangerous, and surprising in every detail. I delighted in the varieties of foliage rising above the ridged topography of this gorgeous land, the rivers and waterfalls, and the unexpected clumps of human culture that sprang out of dense bush.

On one particular trip, I travelled with three American medics— interns or residents, taking time off—and two Malays who served as guides, translators, and helpers. The Americans were Peace Corps volunteers or USAID workers, signed on for two or three years. They were upbeat and friendly. Somehow they accepted my unusual status, or they just didn't care—young guy here in Malaysia with his doctor dad, coming along for the ride. Let him come. We travelled as one seamless unit, with no attention to race or status, by jeep when there were passable roads and on foot when the roads disappeared. In fact there were no roads of any sort where we were going. We were on our own. Our stopping points were Malay and Orang Asli villages scattered along waterways or headlands in the jungle, villages that had somehow been earmarked for medical intervention, immunization, replenishing supplies of drugs, vitamins, condoms, or whatever else was needed. Some of these villages had received medical intervention in the past; some had never had any kind of health care.

A six-hour trek led us to a small, isolated Malay village surrounded by jungle. Our guides must have been known to some of the villagers, or vouched for by the local *bohmo*—a title that conveyed two parts shaman to one part elder statesman—because we were welcomed without hoopla or alarm. I was almost dead with fatigue, but I looked up when I heard the sound of children playing. I shook the sweat out of my eyes and saw that we were approaching a cluster of huts arranged around a dusty common area. Each hut consisted of two or three rooms, separated by doorless doorframes in walls made of bamboo slats, rising above a bamboo floor and covered by a sloped grassy roof. The huts stood, surprisingly steadily, on stilts, so that their floors were just above eye level to someone standing outside on the ground, accessible by a wooden stairway supported by bricks or concrete blocks. These huts were open to the world. They had no front door. They were on stilts to protect the occupants from wild animals or from floods, depending on whom you talked to. But wind, rain, and the curious eyes of neighbours could not be kept out.

As we approached, the common area resolved into a flurry of motion. Children of all sizes trotted around in small herds, laughing and shrieking, chasing and being caught, the tiny ones hoisted on the hips of big girls, maybe their sisters. It was marvellous to see them converging, diverging, like flocks forming and dissolving in front of me. And then there were the adults, suddenly appearing in places that had been empty a moment before. Standing in twos and threes, staring, keeping a safe distance, but inescapably curious. Introductions were made, and then we were ushered into someone's front room, into almost total darkness until my eyes got adjusted. But I could barely make sense of what I saw. The furnishings and accoutrements spanned cultures, if not civilizations, crossed decades and continents, with no sensible pattern, no overarching logic, to any of it. What I noticed were articles of cookware, clothes and fabrics, cigarette lighters, electrical gadgets that looked like they hadn't worked for years—nothing resembling electricity to feed them. No wall sockets. In fact, no walls. So why a toaster? Why an alarm clock when there were no appointments to keep? These cast-offs from the modern world filled holes in a mosaic for which they were never intended. But most of what looked useful came, not from a store or the back of a car, but from the forest. Utensils, rough-hewn seats and small tables, bowls woven from reeds, wooden spoons, ladles, and plates. I gazed about in wonder, smiling, sweating.

After being served tea, the other guys and I were separately escorted to various huts. I really didn't understand what was going on, but I was content to follow. A boy of ten or so took me to the house of the village doctor, a guy named Terry, where it seemed I'd be staying for a couple of nights. He pulled me by the hand, looking up into my face every few steps. His face was wide open, disarmingly present and vulnerable, his anxiety palpable but apparently overcome by curiosity. I don't think he'd ever seen a white person up close, and he'd surely never touched one. He led me to a hut a bit set off from the rest, and indicated with his eyes that this was our destination. But he did not let go of my hand. He stood there staring up at me, without speaking or smiling. I felt strangely flattered. But I eventually shook myself free and climbed the stairs. There was no one inside. I sat down on the top step and sank into a state of lassitude—maybe this was the tempo of the place, this speed somewhere between stop and slow. For a long time I just sat. Maybe I dozed off, because I was startled by a voice coming from behind me.

"Hi." Where was he when I'd arrived? When had he gotten here?

"Hi," I answered. I looked into eyes that combined warmth and humour with some hint of aloofness. As if to say, "You're welcome here, but not for long."

"Are you Terry?"

"Yes, and you must be Marc. I'll show you around." I couldn't tell if he was joking. There wasn't any "around" to show.

"This is where you'll be sleeping," he said, pointing to a spot on the floor. When I looked more closely, I saw a mat woven of very thin strands—reeds or bamboo.

"Thanks," I said. "Where do you sleep?" And why do I care?

"Right here, next door." He pointed to a space a metre or so away from my mat. Indeed, there was another mat there, almost imperceptible in the gloom. "But I don't actually sleep," he added.

"I don't sleep much, either."

"No, I don't sleep at all." He glanced at me. "I meditate." I found this interesting, appealing, and highly implausible. Okay, maybe you meditate, but everyone needs sleep.

I asked him why he lived here, so far from anything, and he explained that this was his sabbatical from life, his chance to be far enough away from the world he came from to invent himself on his own terms.

"You get formed by others' expectations from the time you're a baby," he said. "You never come to discover who you really are, or, I should say, who you *could* be."

"Like . . . what?"

"Like completely conscious. It's possible, you know. Some people call it cosmic consciousness. Whatever. There's a lot you can learn from looking inside, without worrying about how you're coming across to others or what you're planning for dinner."

"But where do you start?" I was testing him more than anything. Was he a pro? Living in Berkeley, I had heard quite a lot about meditation and the provocative goal of cosmic consciousness. I thought that acid trips provided moments of awareness so profound and extensive that you got a taste of it. But that's all I'd ever got: a taste.

"I'll show you a technique I'm working on," he said. "Tomorrow." And that seemed to be the end of it.

We feasted in another hut that night, drank local beer, then came back to Terry's place to sleep. We lay there, listening to the jungle sounds. Terry wasn't very talkative. I finally dozed off and, sure enough, every time I woke up and looked around, he was sitting against a wall with his eyes closed. Maybe you just sleep sitting up, I thought.

The next day I did very little. I explored, helped out a bit, watched the world go by. Took a shower in the woods, using a bucket cut in half to draw water from a well and splash it over me. I was nervous, alone in the woods. There were so many kinds of insects. I hung around in Terry's hut that afternoon, sheltered from the heat. I sat on the steps for hours, looking out on the village scene. Kids kept coming to gawk at me, but that was okay.

That morning Terry had left me with a manual, neatly penned in a spiral notebook. He said it was a step-by-step do-it-yourself guide to developing cosmic consciousness. He said he was still working on it, but the first ten or so lessons were complete. And I wouldn't get farther than that. I picked it up and started browsing. Lesson 1 was all about breathing—conscious breathing. Listen to every breath, hear yourself inhale, hear yourself exhale, feel what your body is doing when it takes in air: because it takes in a lot, not just air. And when it expels air, it gets rid of a lot: feelings as well as air. Take in, send out, take in, send out. That's the basic rhythm of *being*. Lesson 2 got you to attend to your thoughts. Where did they originate? How did they connect, one to the next? Was there any space between them? What was going on in that space? I began to see where he was going with this. But it was hard to believe that some guy in the Malay jungle had found a regimen that would actually lead to that exalted state. The Buddhists, yogis, chanters, and meditators I'd met seemed hardly to have gotten to first base. Was Terry any farther along? If it worked, I wanted some. I wanted to stop trying to change the way I felt, obsessed as I was with drugs, and just accept being in the present, just here, just myself. So Terry's plan seemed attractive to me, whether it was workable or not. His goal was really the same as Krishnamurti's: Be aware. Pay attention to thoughts and feelings. Understand what's going on in your head. That's the path to peace.

Now it's night-time, and Terry isn't the only one awake. There's a little party taking shape. I follow four white guys away from the village, down a path, through clotting darkness that fades finally to a complete absence of light, the stars disappearing beneath the treetop canopy and the firelight fading to blackness behind us. The air is dense with the sounds of unseen life. Creepy. The first guy has a powerful flashlight, and so does the last guy, the one just behind me. I tell myself they know what they're doing. I'm safe as long as I stay in line. But I also tell myself that I can't see the leeches and spiders. Probably just as well.

We arrive at a small, square, one-room hut, perched on a concrete sleeve, a place for storing equipment of some sort, now unused. We arrange ourselves around a tiny paraffin lamp, sitting on the folds of old parachutes that shift subtly with our movements. I was told this is to be a nitrous oxide party. Come and get high. That's an invitation I've rarely refused. But nitrous oxide? That's kind of a cheap high, isn't it? Not that I've ever tried it. And I'm surprised to find Terry here. Doesn't seem like his kind of thing.

Nitrous oxide isn't a common drug of abuse. Its more benign identity is laughing gas, though it doesn't always make people laugh, and it is more often found at the dentist's office, or in teenage stoner parties, or Grateful Dead concerts than in the jungles of southeast Asia. It's a simple drug: the simplest form of general anaesthetic. A compound of two of the most common elements on earth: nitrogen and oxygen. And what it does is quite simple too: it makes you go to sleep. But its clinical usefulness and its attractiveness as a party drug come from the controlled phasing of consciousness. The longer you inhale it, the closer you move to that mysterious threshold between being awake and being asleep. You can stop before going all the way. In fact there's a zone in which you're not conscious enough to feel pain or anxiety or perhaps even selfhood, but you're not entirely unconscious either. You remain a denizen of this world, rooted on this side of the line between night and day. But without quite being here. The most intimate feature of existence—consciousness itself—somehow goes missing for a while.

While watching two of the guys unwrap a large steel canister, the size and shape of a scuba tank, I pester Terry with questions. "Is this a serious thing? I mean, can you use it to explore your consciousness?"

"I guess you'll find out."

"Well have you ever used it for such a noble purpose? Has anyone?"

"William James, for one. Do you know who he was?"

James was a great American psychologist from the turn of the twentieth century. I read a selection of his writings for Psych 100 two years before, but nitrous oxide wasn't brought up, not surprisingly. "I know he was interested in consciousness, but I didn't know he was a druggie. Like Freud and cocaine?"

"He wasn't just interested in consciousness. He was completely obsessed with it," Terry replies. "He thought nitrous oxide would get him a glimpse of absolute reality by fine-tuning his consciousness, bringing it to a point."

"How do you know he wasn't just trying to get high?"

"Getting high for him meant understanding the mind, which meant understanding consciousness. That's the greatest high imaginable, don't you think?"

James's use of the drug was powered by the most respectable of motives: his fascination with the fragile, flitting nature of consciousness. A delicate constellation of cognitive bits that come together to form a whole, a porous whole, like a cloud or a flock of birds. Nitrous oxide was freely available and legal then, as it is now. People have used it for both anaesthesia and recreation for at least two centuries. But James's flirtations with the drug led to some intriguing philosophizing about the nature of consciousness and its access to reality:

> [T]he keynote of the experience is the tremendously exciting sense of an intense metaphysical illumination. Truth lies open to the view in depth beneath depth of almost blinding evidence. The mind sees all the logical relations of being with an apparent subtlety and instantaneity to which its normal consciousness offers no parallel; only as sobriety returns, the feeling of insight fades, and one is left staring vacantly at a few disjointed words and phrases, as one stares at a cadaverous-looking snow-peak from which the sunset glow has just fled . . .

Sounds good in theory. Except for the comedown. But like James, my comrades in arms, these oddballs gathered in a hut in the jungle at night, must be after that moment of augmented awareness, of seeing reality as it really is rather than glimpsing its reflection as it flits past. A macroscopic, supercharged vision of the kind of awareness that Terry is also pursuing. Yet Terry seems to approach the problem from the opposite direction, from a hard-fought, disciplined practice of mental self-control. For me, it's just another high. I admit it. In fact I'm delighted to find this drug thrust on me, especially because it's one I've never taken.

I sit here with the others, silent, senses wide open, jungle noises flooding in from windows without glass, from every crack in the wall, a soft commingling of sound. I watch the apparatus being assembled, some unearthly do-it-yourself kit. Through a complicated-looking valve, they attach a sectioned rubber hose ending in a gas-mask-like piece that fits over the mouth and nose. Someone turns a knob and a low hiss fills the room. The hiss gives the jungle sounds an ominous quality. The sound of an approaching waterfall.

Following unheard negotiations, a guy I don't know—Harold or something—takes up the mask and holds it to his face. He motions with the other hand. The valve is opened more widely and the hiss grows louder. Harold closes his eyes and breathes, once, twice, three times. His eyes flutter and his head begins to nod, seemingly without control, then he rips the mask away. He is falling backward. I watch, stunned. Two guys catch him and ease him down into the folds of silk. Harold's face contorts: a loud gush of air comes out of him, part moan, part sigh, then disintegrates into laughter as his eyes flutter open. "I'm the tank now," he says nonsensically. He looks around the room with great effort, as if trying to focus, his eyes barely registering the faces regarding him, the hands holding him. And then comes his voice again, sluggish and dense: "Wow man! Thas what . . . I can tell you . . . everything . . . the Heart of Darkness, we are it! You guys . . . I *know* you guys. Your hate, your folly, not your f-fault, pure eff . . . effort of life . . . here . . . we . . . just . . . gotta do it, see it. I'm not going anymore." A long moment of silence as he regards some invisible drama. "Wow, what a trip. Nice . . . nice . . . okay . . ." and he passes the mask to the guy beside him.

This one keeps puffing until he falls over sideways, and the first sign that he's still alive is a horrendous cascade of snorts and giggles. He doesn't say much. Just a few exhortations, a few expletives. He looks pretty happy. I'm not worried about him.

Then it's my turn.

I fit the mask around my nose and mouth. Anxiety creeps up my spine. Someone is helping me, but I can't see who it is. A gentle voice says, "Breathe, just breathe." And I do. The gas tastes chemical, or is that the taste of the tank? Something burns at the back of my throat. I let out my lungful, and the world changes. I didn't expect it to come on so fast. Time changes. It slows down, but not smoothly: it begins to lurch. The exhale takes longer than it should, than it possibly could. It just keeps going and I listen, dumbfounded. Perspective glides and shifts. I'm standing on a rise and looking out over this vista of consciousness receding. Then I breathe in again, and this time the gas is a flood of heavy liquid, filling every crevice. Not pleasurable, just shocking. I forget what I was doing but again that gentle voice: "Let it out." And I do. This time the exhale lasts so long that it isn't an event but an episode, a backdrop. The words from Terry's manual are bubbles, expanding as they approach the surface. Then I'm inside one bubble, hopping to the next. I'm inside the manual, which has become a drama with a secret door that I've only now discovered. Here the profundity of breathing is revealed. Breathing out, all of me rushing out through my mouth and nose into the room. Breathing in the universe. It isn't just breath, it's *me*. It's the essence of me. Again I come back to some recognition of myself, as though my existence is constituted by the act of remembering. And that seems so terribly profound. The manual is the lock; the gas is the key. I am the arrow of wisdom, released by the gas, knifing through the pages, past breathing, past thinking, past the space between the thoughts, a space opening up like a galactic cloud. Again and again I awaken, but I don't remember being asleep. I get terribly excited by the magnificence of it all. Then my excitement rises over the rim, transforming into joy.

I lower the mask, trying to understand what's happening. The guys are smiling at me, nodding in approval. But now that I see it, the whole scene is outside of me, far away from me. Their movements and words are distinct, but they belong to a world I'm not part of. The room and its occupants melt back from the island of my existence. Voices come to me vastly distorted, echoing in a canyon of frozen time. I see them as well as hear them: long, undulating tendrils of disconnected meaning. I perceive the act of perception, but it's all wrong. Every sensory event is an absurdly prolonged after-image. I lean back into the silk folds. The tiny light in the centre of the room is half a city block away. I tell myself that this is a trick. If I concentrate, I can close the gap. It's just that I can't concentrate—not on the light, not on the voices, not on the movements or facial expressions. This isn't cosmic consciousness. This is intoxication. My awareness swings through a dispersing crowd of fractured images, making arbitrary sense from arbitrary nonsense. And now the ember of me is swimming up to the gauzy surface of wakefulness, toward clarity. Distorted but familiar voices, pulsating jellyfish, slowly becoming human, welcoming me back.

The effects of nitrous oxide result from the blockade of NMDA receptors, a key mechanism of the dissociative drugs already described. Like its more powerful and longer-lasting cousins, dextromethorphan and PCP, nitrous oxide blocks the access of glutamate molecules to NMDA receptors, disrupting their work of making connections. And the result, again, is a breakdown in the *sense* the cortex routinely fashions by connecting the dots of sensory perception and working memory. As with other dissociatives, other NMDA antagonists, our best guess as to reality goes missing, and *meaning* rushes onto the stage without it. Meaning guided now by associations, leaking in from the limbic system, filling the gaps. But nitrous oxide does something else: it mutes the action of acetylcholine, the neuromodulator responsible for normal alertness, for staying awake. If there is one neurochemical that spells consciousness—the state of being present, aware, alert, sentient—it's acetylcholine. And without acetylcholine to keep you awake, limbic imagery has an open ticket, a free pass to rush onstage and devise little skits about what's going on in the world. The result is unmoored meaning. It looks like truth. It looked like truth to William James and the explorers who followed him. It *feels* like truth. It has that profound gushing certainty that comes when the clouds disperse and the landscape is revealed. But its credentials are highly suspect.

Without the cognitive cohesion of NMDA traffic control, encoding a reality that's sensible, meaning has to come from somewhere else. And like other dissociatives, nitrous oxide gathers meaning from the limbic system and regions of cortex closely joined with it—the orbitofrontal cortex, the temporal lobes, and the anterior cingulate cortex, for example. But unlike other dissociatives, it provides an extra helping of artistic licence by sending me halfway to sleep. It's quick. I only go there for a moment. But that moment is a creative one. Because this state of dimmed consciousness is the place where dreams are grown. It's the state just before being fully asleep, called hypnagogic. The regions responsible for creating meaning are now on their own, detached not only from cortical sense but also from that seamless flow of awareness we refer to as *being awake*. My mind is still making meaning, but this meaning is composed of cascading associations, the likeness of things, cemented by emotion, while my brain stem—the part that does the grunt work—is only half awake, responding sluggishly, with the blinds drawn. There's no way I'm going to achieve any semblance of reality. Rather, I'm throwing together big chunks of perception arranged by the creative logic of dreams.

Dreams have that unique capacity to seem entirely real, to achieve the most vivid representations, plausibly sequenced, and yet be completely invented. EEG (brain-wave) recordings show very little difference between the waking cortex and the sleeping cortex, at least in the first stage of each sleep cycle—dubbed REM, for "rapid eye movement"—when dreams are most vivid. In fact one of the only differences found in the cortex during REM sleep is reduced activation in the dorsal-lateral prefrontal region, the executive suite responsible for working memory. The one thing we don't do well in dreams is keep track of the continuity of events—a prime feature of working memory. That's why we can take the elevator up to our office and find ourselves in high school or meeting our mother the next moment. There are more profound differences farther down, in the subcortical regions, the hypothalamus and brain stem. In these more primal zones, a reduction in acetylcholine (and other neuro-chemicals) pulls metal shutters closed tight across the avenues between body and mind. Sense organs no longer feed input to the sensory cortex, and the motor cortex no longer sends output to the muscles. So we perceive without seeing or hearing what's present, and we act in our minds while our bodies lie still. The mental world remains detailed and precise, but its contents are entirely invented.

In fact, the momentary tableaus in dreams are more coherent than the experience of what we call reality. That's because they are so highly processed. Meaning filtered from experience, filtered and refiltered, night after night. From that meaning comes an alternate universe of sense—the sense of dreamscapes—which gets implanted in the hippocampus and fed back to meaning. Now, with nitrous oxide percolating through my neurons, with working memory van-quished, the distances between words in a sentence, between gazes in a dialogue, between this thought and the one just before it are chris-tened with mythical meaning. This is the magnificent clarity that William James celebrated: insight grasped in all its cohesion, its remark-able comprehensiveness. Except that it's all a dream! Converging from particles of perception, unfettered by working memory, unchecked by the dorsal machinery of self-monitoring and decision-making, this contrived reality is the genie released from the canister. And finally, after the swirling images retreat back into the lamp, there's nothing left at all. Nothing but the sad, vacant after-image mourned by James. And I have to wonder whether that dream of reality bears any resemblance to actual reality. Whether this or any other dissociative drug experience is worth anything more than the thrill of its momentary vividness, its compelling strangeness. Or whether it's a worthless counterfeit, all the more creepy because it seems, for a short time, so profoundly true.

Perception uncurls slowly while the noises in the room grow louder and more intelligible. The rhythmic rush of my breathing comes back to me. I feel myself swaying, dipping, dissolving, and reforming. The faces of those around me are consolidating, one by one—pools of reality condensing in the dim light. The movements of the others, the sight of the tank being withdrawn from me, back to the composite of arms, hands, faces, that direct its path, this coffer, given and received. Now I see what's going on around me. And I think back to Terry's recipe for cosmic consciousness. I imagined I was inside his book a few moments ago. I thought nitrous oxide might be a kind of shortcut, or at least a stage light cast on the noble pursuit he describes in those pages. But now skepticism fills the gap. This was, after all, just a dream.

To compensate for reality lost, the dissociative drugs give you access to a fun house of disturbed consciousness, the careening lunacy of dreaming while awake. And then you can sit back and laugh about it. It wasn't that different from getting drunk. Tennant and I had laughed at the meltdown of consciousness in much the same way, back in the forests of Massachusetts all those years ago. And now, by dimming the lights of normal awareness, nitrous oxide also made way for a secondary illumination, an artificial brightness, intriguing because it smacked of freedom from the mundane. I was okay with this consolation prize, but I don't think William James would have been.

———

Terry and I sat at the top of the steps and watched the stars fade as dawn approached. If he didn't have to sleep, well, then neither did I. But his stillness frustrated me. What was he thinking? Was it of value to him? I had nothing but a mild headache, a lot of fatigue, and a sense of "so what?"

"What happened, Terry?" He didn't respond for a while.

"Come on," I persisted. "What . . . where did you go? Anyplace special?"

"Why ask me?" he finally replied. "You did it too."

"But, I mean, was it important? Was it special? That place, where we went—okay, where *I* went?"

"Is this special?" he replied laconically. "Sitting here with the stars slowly vanishing? With the jungle noises all around us?"

"But . . ." I was getting tired of his silences, interspersed by double-talk. Yeah, I know, consciousness is grand.

"You're really taken with drugs," he went on. "But tonight . . . that was just a diversion. Unless it taught you something."

"Something . . . like . . . ?"

Again he was silent for a while. Then he said, "Nitrous oxide doesn't give you consciousness. It takes it away." And with that, he grinned at me. "Just bonk yourself on the head with a baseball bat if you want to lose consciousness."

"Well, then, why did you do it?"

"Just to be friendly. And I don't mind bonking myself now and then. But look . . . don't make a habit of it."

I said goodnight and lay down on my mat. I felt that familiar sense of wrongness, a kind of emotional rash, an emerging discomfort brought about by too much thinking, doubting myself, doubting what I was doing. Maybe Terry couldn't see it, but consciousness wasn't always such a blessing.

—

The next day, the six of us walked away, inland along the path, after saying goodbye to Terry and his village forever. Once the village disappeared, we were alone in the jungle. We walked from morning till night. The contours of the land and the rainforest canopy were extreme, as though painted by a hyperrealist. The land undulated in a series of enormous ridges with deep ravines between them. But the weirdest thing was the path. Along the lowlands, our path hopped on and off a huge pipeline, carrying . . . who knows what? Probably water, maybe oil. The pipe was about two metres in diameter, it seemed to go on forever, and we walked along it for up to half an hour at a time. Then we'd lose it for a while, then climb back on top for another kilometre or two. The top of the pipe was a thin ribbon, slimy with moss, and the risk of falling was constant. That wasn't a problem where the pipe sat on the ground. But the land would fall away suddenly when the pipeline crossed a ravine. I'd look down and find myself thirty or forty metres above the earth. My heart would freeze, but I had to keep walking. There were guys right behind me. Sometimes the path went straight up the sides of massive hills. In places it was so steep that we used the roots of the trees as a ladder. These roots jutted out of the earth because the hillside was too precipitous to hold them.

At one point we pulled ourselves up to a plateau, gently wooded like the sort of deciduous glade you might find in New England. I stood there on this thankfully level spot, completely exhausted, sweating heavily. Then I was stung by a bee. It wasn't a terrible sting—just a regular bee sting. But three or four others buzzed around my head, and I must have lost it because I yelled "Bees!!" and began running in circles. My cry was heard by the others, and each ran off in a different direction. Minutes went by, and I realized with increasing alarm that there was no sound, no motion, anywhere. I stood in the glade and listened. I was alone, and that was far worse than a few bees.

I wandered around for ten minutes, calling to the others. They finally reappeared, more or less together, approaching me with wide eyes.

"A snake! You saw a snake, eh? How big was it? A cobra or a boa?"

"Nah, it was a wild boar, right? You don't usually see them in the day, but sometimes you do. How'd you scare him off?"

I soon realized that each of them had come up with his own interpretation. All they heard was me yelling, and all they saw was me stomping directionless through the bush. With increasing embarrassment I informed them that I had been stung by a bee. Only a bee. They laughed at that. But nobody saw it quite the way I did. It hit me that this is how people perceive things, normally: they make guesses, and those guesses *are* their reality. That's as close as we come.

For the next two nights we slept in simpler villages. The last night we ate plates of rice with scraps of vegetables, in a hut so full of smoke I could hardly breathe. The smoke came from an open fire, somehow kept alive on a stone platform in the middle of the room. Bits of food fell through the cracks in the floor, providing nourishment for the dogs and chickens who lived below. Our hosts were Orang Asli, not Malays. They did not speak Malay, and only one of us, a Malay guide familiar with the region, knew enough of their dialogue to make conversation. Regardless, they sat with us for hours, exchanging small talk through our interpreter, smiling and laughing. I knew I should be intrigued, but I was bored and restless. They seemed to relish us, to draw something special from our novelty. I never knew quite what.

The following day was our last. We were finally approaching the backbone of the Malay Peninsula, a ridge that ran continuously from the top to the bottom of the country, with smaller ridges radiating out on both sides, tumbling down toward the Malacca Strait in the west and the South China Sea in the east. Here at the summit were the Cameron Highlands, a paradise of cool breezes blowing across hilltops at least a thousand metres above sea level. The hills were covered with tea—miles and miles of minty green shrubs—interspersed by occasional clusters of giant trees, each with its retinue of primary growth. And plantations. Big houses. Running water. Good food . . .

Just before reaching the Highlands, almost desperate now for a shower and a normal meal, I met a pair of Orang Asli who looked like they'd just stepped out of a time machine. I was ahead of the others, so I was the first to see them: a man and a boy of ten or twelve—father and son, I was sure. They stood so still that I almost bumped into them, and my first reaction was fear. The man held a blowgun by his side. It was as long as he was tall, and I knew it was equipped with a poison-tipped dart. He held it casually, upright beside him. He wore only a loincloth. The boy was completely naked. We gaped at each other, maybe for just a minute but it seemed much longer. The man looked strong, confident, and proud. Not the kind of proud that comes from collected accomplishments, but the kind that comes from being completely at home in the world. His smile was magnificent. He seemed to revel in this unfathomable moment. There was nothing he needed to say or do. But the boy's expression and stance were even more remarkable. He regarded me with a face so open, so unclouded, that it seemed to lie outside the repertoire of the human. His eyes were a window between his body and the world outside him, uninterrupted by the opacity of self. Not an atom of self-consciousness, not a hint of anxiety, no shyness, no attempt to please. For days I tried to understand what I had seen in this boy, and bit by bit it came to me: he knew himself instinctively, without a self-image to maintain and adjust, without norms or standards by which to evaluate himself. He felt exactly what it was like to be at home in himself. And for this I envied him enormously. Because, no matter how hard I tried, and despite my additional years, I couldn't find myself, couldn't know myself, not like that. All I could find was a collection of evaluations.

The boy stood completely still, with his father's hand on his shoulder. There was no flinching away in anxiety, no concern that he would do the wrong thing and shatter the delicate father-son détente. No contracting in shame, because the father knew him and accepted him completely. No concern about being too strong, because there was no way it would be taken as a challenge. No fear of being too weak, because his father, his family, and his tribe were there to protect him. These were my conclusions, and maybe they were etched partway between rational conjecture and wishful thinking. But beyond envy, the experience gave me a sense of optimism. Watching that almost-man standing on the path near his home, and reflecting later on what I had seen in his posture and his face, I was left feeling amazed and hopeful. It was possible to be wide open and unafraid in this world. It was at least possible.

THE OPIUM FIELDS

B y the time I left KL the following April, I was really ready to go. Life with Dad and Connie had become suffocating. I had little knowledge of what I was holding inside, but I knew I had to let it out before I burst. Like my exodus from Tabor, to the promised land of California, I was itching to break free of what felt like a prison. Benign and beautiful in many ways, but still a place where I had to follow the rules and be on my best behaviour. My efforts to be good, or at least pretend to be good, had exhausted me. I wanted nothing more than to stop trying. I wanted to travel, to explore, to experience. And much of what I wanted to experience was the cornucopia of drugs that I was sure awaited me.

I knew that the constriction I was trying to escape didn't come from Dad and Connie; it came from me. And my attempts to be good, to be strong and grown-up and healthy, were not directed only at them. They were also designed to please my mother. I tried to make sense of it. I studied the problem at night, in my room, using my journals as a medium. Even though Mom was on the other side of the world, I felt her gaze always. She seemed to me a strict, even tyrannical, therapist, regarding me analytically from the back of my head. She would arrive at a diagnosis, from my behaviour, my trips to buy chlorodyne at the pharmacy, my continued pilfering of drugs from Gombak and from Dad, and from my thoughts, which I imagined she detected through some psychic conduit. Yes, a diagnosis: I was screwed up. I was weak. I was not a man. Mom's gaze felt like a constant, invasive pressure. And I saw that I could not stop feeling it. That was part of my sickness: my dependence on the approval of others, and her in particular. Her evaluation had always been my yardstick. No, I couldn't stop feeling the pressure. But by turning up my consciousness, that fragile entity we had tried to isolate with nitrous oxide, and by focusing the lens of awareness unflinchingly on myself, I could perhaps grow in spite of it. I could perhaps wave goodbye to the wrongness revealed in her gaze as I moved beyond it, into manhood, on my own.

This was not so easily done. I took one train all the way to Bangkok. For two days and one night, I watched spellbound as an ever-changing mosaic of green, dotted with bright pinpoints of colour, passed my window. I slept fitfully, slouched over in my seat, but with each awakening I felt more myself. I stood on the little platform at the intersection between the cars and let the wind ruffle my hair. All the way up the long, stringy country that was Malaysia, up into the entrails of Thailand, hanging down to merge with it. The border region was given over to guerilla skirmishes and other dangers. Then we were past it. We had emerged from a land dominated by Islam to the mecca of Buddhism. And on up to Bangkok.

Bangkok was enormous—and enormously ugly. Except for the temples, which I admired on my long solitary walks through the city. I got a little lonely, especially at night. So I was glad to meet up with other Americans and find myself sitting around a bong filled with hash of considerable potency. I hung loose with these fellow travellers, druggies, freaks, hippies, whatever you called them, for a few days, sleeping on someone's sofa, talking about all the places I had to see. I was feeling pretty good. And during this period, thanks to my new associates, I learned about the route of young travellers that extended across Asia, from Bali to Kathmandu.

In the early seventies there was a well-known path that passed through Indonesia, Singapore, Bangkok, Vientiane, Chiang Mai (a city in northern Thailand), and Burma, then emptied like a sewer into India, where it branched off to Calcutta, Varanasi (Benares), and Goa, or north to Darjeeling and Kathmandu. This road was travelled, was defined, by hippies of various stripes, clear-eyed idealists or burned-out druggies, seekers of truth, religious nuts, New Age explorers of meditation, macrobiotics, or yoga, budding Buddhists, and long-haired white saddhus, decked out in the local garb. This was where the young and displaced from the U.S., Canada, Britain, Germany, France, and Australia went to find adventures—the intrapsychic variety supplied by drugs and the geographical and cultural variety found in every town and village. These kids weren't all bad. Not at all. Many of them respected the people whose lands they traversed. Many tried to imitate customs that seemed foreign and strange and yet so attractive, so refreshingly sensitive in contrast to the coarse Western habits they'd left behind. But some wore their origins on their sleeves. Some were stupid and rude, loud, self-centred, and disrespectful. I was determined to be a gracious visitor and to leave nothing but good vibes in my wake. I studied French for many hours on the train to Laos, so that I could communicate with the Laotians. But I was not so respectful as to bypass the local harvests of opiates. These drugs both lubricated the economies of Thailand and Laos and destroyed the delicate fibres of their cultures. I didn't realize it at first, but with each stop I saw a little more of the harm that came with drugs, with the places that served them and the people who pursued them.

Vientiane, the capital of Laos, was covered in dust and sick with poverty and corruption. A dying city. Even the trees looked dead: dirty, reddish, mummified palm leaves hanging motionless in the baked air. The temples were tawdry and dull compared to the magnificent palaces of Bangkok. And the boom-boom of artillery could be heard on and off in the distance, day and night, from some frontier of the Vietnam War. I arrived at a cheap hotel and asked around. I wanted to find some serious narcotics tonight. Yes, tonight. Ideally, heroin. And I was told where to go and whom to speak to. I was in a bad mood. A beggar had approached me while I'd sat at a café that afternoon. He'd put his hands out for money and I reached into my pocket. But then he lowered himself to his knees. He must have thought I was going to refuse him and that he needed this extra supplication. He mumbled some incantation over and over, the way beggars in the East often did. But my reaction was extreme: it was a wave of disgust and anger. "Get up!" I yelled at him in English. Can't you see I'm going to pay you? Why do you have to get down on your knees, like an animal? Don't you have any pride?! My sense of propriety was badly jolted. How dare he be so poor, so abject, so conspicuously beneath me? I just wanted him to get away from me, and I dug all the more desperately for change. I hated him for making me feel so ashamed, so exposed by the vast inequity that defined our relationship.

At sunset, I walked over to a compound in the centre of town. I wasn't going to shoot anything. That was a promise I'd made to myself on the train from Kuala Lumpur. Because I was afraid. I knew that if I started shooting narcotics in southeast Asia, there was a good chance I'd die there. No, I wasn't going to shoot anything; I wanted to buy a supply of heroin and snort it. The intense midnight blue of that ultimate opiate would wash away the dust I'd accumulated over eight months of stagnation in Malaysia and the dust already settling on me here, as a moneyed hedonist tromping through this troubled land. Someone pointed me to a drug and gambling parlour. I stood by the door, trying to get my bearings, then asked the proprietor where I could buy heroin. He was happy to oblige. He sent for a young man, who came to the door ten minutes later and then led me to another building. He was so thin, his fake blue jeans seemed to hang from his limbs. His smirk implied that he knew who I was. He knew my type. How much did I want? How much would it cost? Because we had very little language in common, it was impossible to understand his chatter. But I gave him five U.S. dollars and he disappeared for a few minutes. Five bucks wasn't much, and I was already prepared to be cheated. I'd try somewhere else if he didn't come back. But he did come back. With a bag of white powder—a shocking quantity of it.

I returned to my hotel room. The plastic baggie was a quarter full of snowy white powder. This was a lot of smack. So how good could it be, at that price? Was it even real? But this was Laos. It was probably manufactured a few blocks away. Be careful, I told myself. I sat at a small table and tapped a tiny pile on top of the paperback I was reading. I rolled up a piece of paper to make a small tube, put one end into my nostril and brought the other end down to the mound of powder. I covered my other nostril and inhaled softly, careful not to take too much. But when I looked down, I saw that the pile was essentially undisturbed. Had I missed it? I was about to lower my homemade tube once again when my guts began to contract. The contours of the room started to darken and go hazy, as though covered with fur, and a familiar sensation, something between spinning and shrinking, engulfed me, pulling me down into myself. I was rushing on heroin. And I'd taken only a few tiny grains into my body. If I had inhaled the rest of the small pile before me, I'd be unconscious by now.

That night I went to a prostitute for the first and last time in my life. I told myself this was to be an evening of the most thorough debauchery, and I was proud of my grown-up bad-boy competence. I'd gone all the way, hadn't I? I was proud until the next morning, when I insisted on walking her home. We walked along a narrow dirt road through forests and gardens too dim to make out. The air was fresh and fragrant, yet everywhere was shadow. I needed her to lead me, but she would not look at me and there was nothing to say. She simply walked, head slightly down, mostly behind me, pulling ahead only when it was necessary to make a turn I would have missed. It was still early light when we got to her door. I prepared to say goodbye, imagining some kind of bond between us, but an older man suddenly appeared and called to her gruffly. His blank, half-opened eyes expressed only greed—*How much did you get?*—and then he turned away, as though she did not exist. On the long walk back to my hotel, I thought about what I had done. She must have been a servant, or a slave, or his daughter, or who knew what? And I had paid into the deal. I had made it work. I felt terrible.

I spent two more days in Vientiane, but there was little to do. So I bought a ticket for a boat trip up the Mekong River. This would be my chance to see the real world, the outside world, up close, where I might be able to glimpse the land and its people now that I'd adjusted my inner landscape to my satisfaction. The night before departing, I sat at the table in my room, loading six or seven tetracycline capsules with heroin and throwing the rest away. The boat trip would end at the Thai border, and I didn't want to imagine how the border authorities would react to a white guy travelling with a bag of heroin. Now, with my heroin-stuffed capsules back in the prescription bottle, I felt confident I'd have enough to last for weeks, but without the danger of getting busted. After eight hours of travel upriver—eight beautiful, mystical hours watching the hills and paddies of southeast Asia go by—the border crossing turned out to be a small town on the Thai bank, the left side. But I could see nothing resembling a checkpoint or customs office. I was pointed upriver and told to keep walking up the path until I got to a small house. Twenty minutes later I was nearly certain I'd missed it. And then, there it was: a tiny house with a sign above the door announcing "Customs." It seemed like a joke. A middle-aged woman answered my third knock. She couldn't be an official. She wore nothing above her waist but a bra, and she seemed almost asleep. She couldn't be . . . but she beckoned me inside, sat herself behind a cluttered desk, and pulled a stamp and a pad from somewhere inside it. I held out my passport and she stamped it with profound apathy.

I spent two or three weeks in Thailand, travelling by train, bus, or whatever came by. Each day brought the unexpected. Each night I settled myself with something familiar: the soothing goodnight kiss of opiates. I met an assortment of people. Some were young and lost like me. Among them I could always find more drugs, which came in handy when my heroin ran out. At sunset one night, I met a guy who took me on the back of his motorbike down dark roads surrounded by invisible fields. He stopped and went into a house without a word to me. He came back—not with a knife, as I'd begun to fear, but with something called "red rock," a cheap derivative of the heroin-making process. We smoked it on a beaten-up sheet of tin foil suspended above his lighter. Then I got back on the motorbike and held on for dear life. I was smashed, and he must have been too. We rode to a large house, lit up by a dozen blazing windows, like a gigantic lantern sitting in a midnight garden. And even at that late hour the house was alive with activity: I saw fifteen or twenty young women— his sisters and cousins?—all of them beautiful, so it seemed, at work in rooms all over the house, making umbrellas. Umbrellas adorned with delicate images of birds and butterflies, in bright pastel shades of yellow, blue, red, green, orange, and pink. I was given a bed in a small room, with a door that shut out the light but not the sound of young women chattering. I lay there breathing air that was fragrant with the scent of sandalwood.

Then came a couple of weeks in Burma. I went to various towns and cities without any particular agenda, planning my route each day according to a conversation with someone yesterday or the day before. I wanted to be present in this world of unfathomable cultures. There was so much to experience here. But my path seemed to be determined by my inner landscape instead. The ongoing search for new highs. I finally bought a one-way ticket aboard a small commercial carrier called Indian Airlines. The plane shuddered to a halt in Calcutta. India definitely seemed to be the end point, at least for now. I would stay here for a while. I thought I would buy a sitar and take it home and really learn how to play it.

But I was in trouble. This dawned on me, bit by bit, while staying at a hotel called the Modern Lodge, a sleazy three-storey model of ugliness in the middle of Calcutta. Here I developed the habit of wearing my sandals in the shower. No way was I going to let my bare feet touch that floor. I slept on my own sheet, and I was constantly on the lookout for cockroaches and bedbugs. It was here that my relationship with opioids stabilized, a marriage finally settling down after the roller-coaster ride of early infatuation. But this took time. My addiction descended like a shroud over a terminally ill patient, rising and falling with each shuddery breath. Some nights I kicked it off in a fit of frustration, a small-scale version of the rage that had gripped me when I'd said goodbye to Jimmy last year in San Francisco. But I really was in trouble this time. I had gotten high nearly every night, not for a week but for two months. The whittling of my orbitofrontal and striatal synapses had gone on so long that, at least for now, there was no flexibility left, no raw material still to carve, no receptivity in these arbiters of value and motivation. Meaning had now crystallized. Nothing else mattered very much. And the net result was that my energy, my enthusiasm, my creativity and optimism pointed in only one direction.

My present drug of choice was opium, the original opiate, and I'd learned to hunt for it in the circuitous alleyways of Calcutta. In those days, raw opium was legal in Calcutta. You could buy it at little government-sponsored wickets, along with your weekly supply of marijuana. Refined opium was illegal, however, and that made sense to me: it was a lot stronger and it was highly addictive. Refined opium seemed to come only from Chinese vendors, and they did not advertise their wares. So the dens where you smoked this stuff were difficult to find, and for some reason they seemed to be located only in the Muslim areas, where twisted alleyways prevailed, hiding secrets you'd never discover if you didn't know where to look. By day, I hung around the Modern Lodge, chatting with travellers passing through on their way to somewhere else. But I quickly got bored talking with Westerners. By evening, I paced the tiny courtyard that passed for a lobby, or else waited for dinner in the crude dining room. Restlessness welled up in me, a rising reservoir of dopamine, reflecting the imminent approach of the only goal, the only show in town, beckoning with the coming of night.

My lengthy stay in Calcutta wasn't planned. I was waiting for my sitar to be finished. Ever since those nights in the Indian Room in Dad's apartment in San Francisco, the luxurious tones of Indian classical music had captured me. After six months of lessons in Malaysia, I'd acquired some of the basics, but I needed a good instrument, not a toy, in order to continue my instruction. And you can't just buy a high-quality sitar off the shelf, like a Selmer clarinet or a Martin guitar. You have to order it, and you have to wait. I'd asked around for a week or so. I kept getting directed to Hiren Roy's shop on Rashbehari Avenue, seemingly the capital of the Calcutta music scene. The never-smiling, perpetually squatting Mr. Hiren Roy was the master of the sitar. First, I came to watch. I spent an entire afternoon, trying to squat without going into spasm, trying to chat casually with Mr. Roy and his minions, in a blend of English and Bengali. One man perpetually sanding the two bridges, supporting twenty strings between them, soon to be glued to the teak face of an almost-finished instrument lying before him. Another man inserting white bone inlays, fiendishly carved, into the indentations in the twenty pegs waiting to receive them. Finally, I told Mr. Roy what I wanted, and he gave me a price, which must have been a lot of money in Calcutta but it wasn't too much for me. Three weeks, I was told, and I would be the proud owner of a Hiren Roy sitar, the best of the best.

I went to the shop a couple of times a week. I was told that's how to get things done here. That's how to speed things up. But there was no sign of my sitar in three weeks. Just more talk, gesticulations, promises. Cups of tea perpetually exchanged. It was discouraging. In six weeks, a sitar was brought to me from the back of the shop. The elaborate inlays were half in place. I was given choices. Did I want the ordinary knobs—the ones with simple flanges curved into a spiral? Or did I want those carved in the shape of lotus blossoms? I kept going back, but I could not seem to accelerate the process beyond some fundamental limit, a physical threshold like the speed of sound or the speed of light. In Calcutta, the speed of sitar was grindingly slow. Now it had been nearly two months, and my sitar was almost ready. It would soon be presented to me. But some of my enthusiasm had worn thin. Slackened by the humidity, stretched over too much time, and impinged on too frequently by my other pursuit, the one that emerged every evening when the sky began to darken. Enthusiasm is the psychological shadow play cast on the screen of the world by dopamine, and my dopamine was no longer freely deployed.

I lie on my bed in the waning afternoon and I try to find a way out of this. Just a few more days, I tell myself, but I hardly care just now. The pride I anticipated in owning a fine sitar seems counterfeit, something I've invented. But I would have that sitar. In a few days, maybe next week. And as far as tonight, it's taken care of: I can already taste it. I get up and pace, and I start to get excited. Dopamine release is narrowed by addiction, specialized, stilted, inaccessible through the ordinary pleasures and pursuits of life, but gushing suddenly when anything associated with the drug comes into awareness. Just a flicker, an image, a word echoing in memory, a rumbling in the gut, that first lurch of withdrawal symptoms, or the scent of hoisin sauce from a Chinese restaurant, eerily similar to the smell of opium bubbling. Yes, dopamine takes over, again this night, when my thoughts turn to the preordained drama about to ensue, beginning with the rickshaws that will congregate here at dusk. Having wasted the day lounging on my bed, or lolling on the ratty furniture in the common room, it's now nearly time. Dopamine's alloy of craving and excitement gathers in my synapses, settling once again in its familiar constellation.

An hour later, dinner is nearly done. The two white-clad Oregonians sitting across from me are impossibly boring, with their talk of unity, the inherent oneness of all things, the manifestation of God in every creature, the universal consciousness that extends to plants and rocks, blah-de-blah-de-blah. For me the only unity is opium, my evening's entertainment, the star of the show. I say goodnight, quickly and vacuously, and I get up to leave. It's become typical for me to disappear after dinner. But I don't want to be rude. I take one last look at the little pools of oddly coloured mush, unnamable vegetables in various states of liquefaction, served on burnished steel plates, arranged on two long tables. Here, young Indian men and Westerners, of various ages and dispositions, many of them down on their luck or their finances, eat together in a suspended cultural experiment. We often eat in silence. I prefer that. But these Oregonians won't let me go without a long, drawn-out goodbye. Which generates a growing tension in my guts, an agonizing tautness as dopamine collects in my synapses: Get out of here and find the rickshaw *wallah*. (Wallah means "guy"—the rickshaw guy.) Find him, before he disappears. He'll expect me about now—at least that's what I've come to believe—and he knows the general vicinity of our destination much better than I do.

I walk out onto the street and sweat starts at once to trickle down my cheeks. Dopamine surges: I have to get there. Dopamine gushes along overextended axons, fattened gutters of chemical enhancement, diverging from the VTA to its three main targets: the ventral striatum, where behaviour is charged, focused, and released; the orbitofrontal cortex (OFC), where it infuses cells devoted to the *value* of this drug; and the amygdala, whose synapses provide a meeting place for the two most important components of associative memory, imagery and emotion. Thanks to the neural reconstruction brought on by addiction, the power of dopamine, the power to convert sensation and memory to value and thrust, is now joined by glutamate, released from the OFC, a messenger of meaning. Those orbitofrontal neurons *know* how good that opium is going to feel. So glutamate leaves the orbitofrontal warehouses like traffic at rush hour. It surges to cells in the amygdala, one synapse away, where it awakens details connected with pleasure and relief—the sight and smell of that rich black tar, and the *feel* of it melting my anxieties. The synaptic glue used by the amygdala is permanent. And its message of sensation and feeling comes surging back to the OFC along an axonal highway where the details converge into a single, global, image: a hunk of pure meaning. Meanwhile, the same glutamate traffic spreading out from my OFC goes back to the VTA, which called it up in the first place, and the VTA responds by exporting even more dopamine, now that it's alert to the immense value now within reach. And that's the feedback cycle at the heart of value: dopamine from the VTA to the OFC, glutamate from the OFC back to the VTA, each feeding the other, round and round (see Figure 4). But the glutamate tide goes to one last target: the royal scribe of addiction—the ventral striatum—and that's where value is converted to action.

So the ventral striatum is suspended between two converging waves: glutamate from the orbitofrontal cortex, carrying value or meaning, and dopamine from the VTA, carrying thrust. Now, thanks to the evolving partnership between my OFC and amygdala, the sensory details are piped through as well. But the wash of dopamine from the VTA amplifies amygdala activity too, providing energy to bolster its detailed reading; and the amygdala sends its message of emotional potency straight to the ventral striatum, arming motivated action with a precise sensory target. I sit at the dinner table, gazing down at my plate, and imagine that liquid pearl of opium dripping from the Chinese woman's skewer. And my ventral striatum says, *That's what I want. That's exactly what I want, and I want it now.* All of which creates a second feedback loop, overlapping the first (see Figure 4). This is the feedback cycle at the heart of thrust, or motivated action. The partnership between my OFC and VTA sends more dopamine to my ventral striatum. More dopamine for my buddy, if you don't mind. Send it direct. And my ventral striatum sends messages back to my VTA through various intermediaries, requesting even more dopamine. That's how craving builds on value or meaning. Yes, value amplifies itself through one feedback loop, and thrust builds through a second one. The net result is craving. Further refined whenever my amygdala is triggered by additional details: the sight or smell or feel of the drug, a step closer than it was before.

These two feedback loops amplify each other. As shown in Figure 4, they are interlocked, because that's how the brain works. But they are even more tightly interlocked in the brain of the addict, in *my* brain, because they shut out competing inputs, alternative meanings. They go round and round, building on themselves, releasing more dopamine and more glutamate in an ever-tightening spiral, a double helix of neural excitation. Producing a state of directed desire: desire that's come to a point, so that nothing else matters. It's not just craving, it's customized craving, narrowly aimed. An ever-tightening spiral that is the outcome of its own recurrent history of synaptic selection.

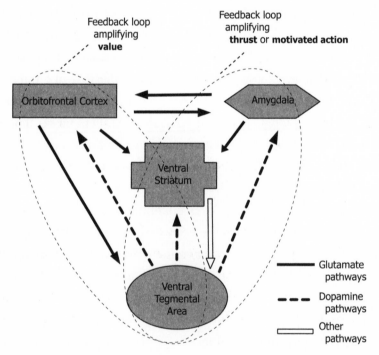

FIGURE 4. Two feedback loops connecting the orbitofrontal cortex, amygdala, ventral tegmental area (VTA), and ventral striatum, through dopamine and glutamate pathways. These loops overlap and fuel one another to create a state of craving or directed motivation centered in the ventral striatum.

All along the pathways joining the OFC and striatum, the *orbito-striatal* routes, the active synapses—the chosen ones—become strengthened with each cycle, while the synapses standing aside, the wallflowers never asked to join the dance, wither and recede. This process of synaptic sculpting goes on in many parts of the brain, and it's how the brain develops in childhood and adolescence. It's how experience moulds the very circuitry of the brain, how memories are formed, how meaning is encoded, how expectancies are laid down. And it's how the addict's brain, fertilized by the emotional potency of repeated drug experiences, develops, like that of a child, but way too fast, way too conclusively: tightening, rigidifying, becoming more caricatured, through its own relentless action. This drug isn't chewing gum. It's evolved for centuries, apace with human civilization, because it feels really good. Because it does the job of pain relief, it provides that incomparable warmth of connection, even better than the internal opioids it mimics. And it keeps me coming back for more. Again and again, night after night. This is the path my brain keeps taking. This is the path *I* keep taking. And the habit grows stronger, the synapses more refined, their connections more crystallized, their options ever narrowed by the unique fatalism of plasticity lost.

That's what's so insidious, so toxic, about addiction. The neural traffic routes get more and more constrained, thanks to the sculpting—the shaping and pruning—of synapses, augmented by dopamine and glutamate, night after night after night, empowered by the emotional potency of the goal—the craving and then the relief, the heights and depths of emotional space. Remember: synapses used are synapses strengthened; they are the ruts in the garden where rainwater flows, forming deeper and deeper troughs. The congealing and narrowing of synaptic traffic, crisscrossing among the OFC, the amygdala, the VTA, and the ventral striatum, leave less and less choice. There are fewer routes to take with each replay of the fundamental story line. Leading to more repetition, less flexibility; more habit, less choice.

The psychological realities of diminished choice and narrowed interests—those well-known attributes of addiction—are precisely paralleled by the neural reality of reduced flexibility in synaptic traffic patterns. But here's the thing: the brain doesn't really parallel the mind. That would be a misnomer, a poetic approximation. It's the other way around: the mind parallels the brain. The way the brain works—the biological laws of synaptic sculpting and neurochemical enhancement, each reinforcing the other—are what constrict the addict's mind, his behaviour, his hopes, his dreams. And I wish the picture were rosier. I wish there were a back door, a side door, a secret opening for mind to rise up over matter. But there isn't. I wish this were just an exercise in biological reductionism, or neuro-scientific chauvinism, but it's not. It's the way things really work.

For me, this night, the craving is bad enough without any help. But having to undergo this interminable conversation with the Oregonians, having to endure their long goodbyes, I've held it back long enough. I've inhibited the impulse, yes, long enough, and now let's get going! The craving is a low-level nuclear pile of physiological arousal, built around a single goal whose acquisition is always just out in front of me. This craving reaches out now to my premotor cortex, the cortical master contractor that moves my legs, my mouth, my hands, now dipping into my pockets to flash a roll of rupees to my rickshaw guy. *Get me there.* My ventral striatum rises up like Ahab's whale, suspended between those two colliding waves: urgency and evaluation, motive and meaning. This whale of mine thrashes with its lunatic single-mindedness. I want it. *Now.*

These neurochemical events are just foreplay, of course; just the anticipation of that rush of relief that comes with the smoke itself.

But after all, foreplay is the thing that's most interesting at this stage of the marriage; the act itself, the primal deed, is relatively routine. The meeting of pipe and lips, of needle and vein, is the same, or nearly the same, every time. It's the means of getting there that provides the drama. So I stand on the sidewalk, sweating, scanning. In air almost solid with soot, still hot and humid at 8 or 9 p.m., double-decker buses trundling by, tilted at bizarre angles, stick figures hanging off the handrails like desperate cartoons. And I see his bony torso, coming toward me. Three other rickshaws approach at the same time, having spotted a white face, each spurred on by the presumptuous advance of the others. I rapidly approach the guy I know, to avoid a confrontation. But for a confused moment, each rickshaw looks exactly the same: a "cab" consisting of a seat wide enough to hold two, maybe three; a folding hood to keep the rain off; a bit of leg room on a sloped platform that ends abruptly. My guy nods at me, and I step up on the small platform beside the large, spoked wheel, then plunk myself in the seat. From the cab, two long poles curve forward symmetrically, so that the "driver" (runner?) can hold one in each hand. He hoists the poles, and this tilts the whole thing back on its wheels, then he begins to trot down the street without looking behind. Where to? I name the closest main street because that's the best I can do. It's beyond the Maidan, that huge park in the centre of Calcutta, beyond the post office and other government buildings, in the Muslim district, where the roads lose all semblance of perpendicularity. But he knows the way by now.

Thanks to his sinewy strength, we arrive in a familiar district in twenty minutes. Serene avenues give way to narrow side streets crowded with small shops, many still open. I see sides of beef, blasphemous in the Hindu neighbourhoods we've come through, now hanging at the fronts of butcher shops, with bearded men wrapped in shapeless white robes sitting in front of them, smoking sweetly scented tobacco in hookahs. Jars and baskets of indefinable food-stuffs, weaving a smell both horrific and compelling. Then, within minutes, the streets disintegrate further, into thin, twisted roads, almost alleyways. This is the part I can never track. I see no street names on the corners of buildings. No numbers. Just shadowy doors or hanging fabrics giving way to unknown worlds.

At first I felt embarrassed to be pulled along on a rickshaw by a guy who looked about half my weight. Mind you, that weight was composed entirely of muscle and sinew. A few times I was drunk or stoned enough to mess with the rules. I'd somehow tricked the driver into getting into the cab or found him sleeping there, curled up, and picked up the handles before he could move, then started pulling *him* along. Each time ended in an explosion of protests, and I would look back smiling—don't you get it? It's just a joke, you see. It's completely arbitrary that you do the pulling and I do the riding, because we're the same, aren't we? Skin colour, status . . . these are fictions. But no, this was worse than the Oregonians' claims about cosmic unity, because I was toying with the funda-mental layout of their livelihoods. The rickshaw wallahs were not amused. You just don't do that. He's the runner. And I'm the pas-senger. And messing around with the roles, because it amused me or satisfied some childish idealism, was an insult I only began to fathom after weeks of living there. The driver-passenger role rela-tionship was built into the way things were. Waving it around debased these men, resulting in a message essentially opposite to what I'd intended. I had demonstrated my power to play with the rules. O bestower of equality! They wouldn't have dared. Now I sit back in the seat. I point and gesture. The wallah, apparently inde-fatigable, takes me this way and that, and it's a thrill to go racing through these exotic streets, especially on our way to Drugland. We finally arrive in a laneway so narrow as to challenge even this minimalist vehicle. And then, miraculously, I find myself a few steps from the opium den I like best.

I walk through the tiny foyer, into a room dense with shadows. A wide shelf—another floor really—juts halfway across the room, covered in some sort of bamboo matting. I take off my sandals at the door, then sit on the edge of the shelf, waiting for my turn. All the clients—all four or five of them—are Indian men. And now the last one is finished. They sit on low benches along the outer wall in unique poses, statuesque parodies of utter relaxation. I would soon be one of them. I step up, nod to the woman, try to smile, and lie down a metre away from her. Our bodies are parallel, facing each other, as though we're about to engage in some form of sex. She gestures to me: come closer. I lie on my side and place my head on a wooden support, a sort of pillow, less than an arm's length from a glass cylinder shaped like a large pear. A flame burns steadily at its centre, its tip ending just below a circular hole. The flame is unwavering.

The woman works with a bead of purified opium on a long pin. Not the raw stuff; this is pure and soft, almost liquid, like thick honey, the size and shape of a large teardrop. She moulds it incessantly, lovingly, just above the flame, until it's a perfect cone, perfectly centred on the pin. Her eyes gleam with indefinable light. She pokes the opium into a hole, obviously shaped to receive it, in the centre of a circular brass chamber attached to a long wooden pipe, a metre in length, a pipe that looks as old as she is. She looks up at me and catches the nakedness of my anticipation. My need, my eagerness, now beginning to elongate, to extrude into a mood of imminent fulfillment. And there is nothing I can do to hide it. The smile that cracks her dry old lips mixes bits of madness and kindness in a blend both shocking and beautiful. I belong to her at this moment: her poison has already infected me, frozen my limbs, paralyzed my heart. I think she may be insane. That smile, complicit and completely unabashed. Or am I the crazy person, dancing this dance with her?

I'm intimately familiar with the steps by now. With both hands she rotates the pipe so the hole is on the bottom, then she positions it above the flame and points the long end at me. The mouth of the bowl is directly above the hole in the glass cylinder, and the tip of the pipe is inches from my face. My only job is to guide this end to my lips. There is no thought of germs. The invisible tip of the flame licks the opium inside the bronze bowl, and in moments it starts to bubble. Then she draws in her breath, a signal for me to do the same. And I do. Harsh smoke tears at the back of my throat, but its flavour is as comforting as anything I know. Dark, rich, and slightly bitter. A cough gathers and almost explodes from my chest, and then it subsides before ever breaking out. Opiates are the key ingredient of strong cough medicines. And this opium quells the cough it has almost unleashed, reducing it to nothing, snuffing out the life force gathered by my body in an effort to expel the poison. My body has surrendered before taking the full measure of this invader. And somehow this trickery, this giving and taking away, this mixture of attack and nurturance, hurt and relief, captures the soul of this drug. The potency of its assault is without equal, but so is its capacity to soothe.

I left Calcutta when my sitar was finished, a few days later. My physical addiction to opium cost me weeks of discomfort, but it passed, thanks in part to the mouthfuls of codeine I swallowed when I got back to Canada for a summer job. My withdrawal symptoms weren't as severe as they'd been following my week of heroin with Jimmy, but they lasted longer. The price tag could have been worse. Then, finally, I was free of opium, but I was still in trouble. My stint of readdiction — more psychological than physical — had laid the groundwork for more misery. Opioids from outside my body had become my ticket to safety. Opioids waited for my day to end, offering a buttress of security against

the dark clouds—boredom, loneliness, shame, circling self-doubt—
that descended anew each morning, made all the worse by the very
fact of my addiction. Made all the more maddening because I saw
myself as a pathetic creature, a fool, so completely obsessed with a
stupid drug that I was impervious to the riot of life, the celebration of
everyday sensation, that even the poorest people on earth were enjoying all around me. I was in trouble because my sickness called for a
cure that only made it worse: more ominous, more dangerous. I was
in trouble because, though I didn't quite get it yet, some part of me
would erode further every time I came into contact with opiates in the
future. My attraction would burgeon with alarming suddenness, my
control would give way, and I would take risks that I couldn't yet
imagine. Opiates made me feel safe and warm, cared for, soothed by
the peptides that my hypothalamus could never make enough of.
They gave me control of the outlay of warmth I felt I needed, and that
I would need all the more to combat the shame and depression they
themselves engendered.

I spent the summer in Toronto, then went back to Berkeley in
September, back to school, and completed a music major: harmony
and counterpoint, Mozart operas and Beethoven symphonies. None
of which contributed to my primary devotion—Indian music—but I
was content. I studied sitar in the summers, with famous musicians
who came to California to escape the furnace that was India and
revel in the adulation of star-eyed sycophants, some of them very
pretty. Three years later, I graduated, then went back to Calcutta for
a nine-month stint of serious instruction. This time I wasn't alone.
Michael came with me, and he studied tabla, the drums that accompany sitar. We practised four or five hours a day, sweat pouring out
of us and then drying under ceiling fans turned up to helicopter
velocity. I went back to my precious opium dens a couple of nights

a week, but the craving was not as intense as before, probably because I had Michael and music to comfort me. But despite its sublime beauty and the self-discipline it brought to my life, this music could lead me nowhere. There just wasn't a market for Indian musicians in Canada or the U.S., and I had no intention of spending my life in India. Besides, I had met a Toronto woman I thought I was in love with. Sharon and I were talking about getting married, and that meant finding a serious profession and picking somewhere in North America to call home. I decided I wanted to be a psychologist. Though I had by no means mastered my own self-realization, and though I continued to take drugs and the risks that went with them, I thought I would flourish in a field focused on the hows and whys of human behaviour. To study the mind still seemed the noblest of pursuits. I particularly wanted to understand emotions and drives and their astonishing power to carve out paths of fulfillment or failure. I thought I was ready. I thought I could rally my intelligence and energy in the service of this goal. And I thought I could do it with a wife to come home to at night.

So I left Calcutta a second time in the late spring of 1976, having started arrangements to get married the coming fall. I'd been accepted at the University of Toronto, where I would complete my undergraduate training in psychology through a second B.A. On the way back to Canada, Sharon and I travelled for two months in Europe, and we argued our way from one country to the next. Some of those arguments revolved around the motorcycle I'd impulsively purchased in England—a motorcycle she ended up pushing, with me on it, to get us started each morning. We should probably have rented a car, as she'd wished. But the arguments wore me down, and there never seemed any resolution to them. That should have suggested a more cautious approach to marriage. But I was determined to make it work. I was committed to becoming the best person

I could be, governed by willpower, noble goals, and the discipline to see them through. I would be impervious to the intrusions of negative emotions; I would not be derailed by self-destructive impulses for once in my life. And yet, the night before our wedding, in October 1976, tying my tie in front of the mirror and drinking scotch with my brother, I knew it was wrong. I knew the marriage was a mistake.

PART FOUR

IN SICKNESS
AND IN HEALTH

13
NIGHT LIFE IN RAT PARK

There was rarely another soul in the subbasement of the Life Sciences building after seven or eight o'clock at night. Just the animals. And me. The first thing I heard when I unlocked the door to the animal wing was the screeching, squealing sounds of the monkeys. Was this cacophony a welcome chorus for me, a reaction to the sound of the door opening, or had it been going on for hours, maybe all day long? Were they suddenly panicked? Or curious? Or maybe even hopeful? I couldn't think about the monkeys, electrodes drilled into their skulls, waiting for the knife. I never ventured into the rooms that housed them or the surgical areas where they received whatever terrible interventions awaited them. My work was on the other side of the corridor, through the first door on the right, where the rats belonging to Dr. Peters lived out their lives. I unlocked the door and found the office just as I'd left it the night before. There was an old sofa to my right as I walked in, a desk against the far wall, a couple of chairs scattered around, and a second door to the left of the desk, opening onto the procedure room, the place where we did our dirty work.

Except that our dirty work didn't involve surgery, and it didn't involve animals with fingers and toes—and brains—that looked a lot like mine. The rats weren't close relatives, but they were my friends, my frisky little partners, helping me to expand the horizons of psychological science, or behavioural science as it's often called. They were the subjects in behavioural experiments conducted by Dr. Peters and his team of post-docs and students. They were my friends, but not my ancestors. And what I did to them wasn't particularly gracious, but there was no cutting, nothing that seemed very cruel, except for the isolation we imposed on them. The next room, adjoining the procedure room, was the cell-block where they lived. It was really a longish corridor with a series of alcoves branching off it. In the second alcove were my rats, the ones Beth — now Dr. Fenlon — used in her research, the research I was helping her conduct for my undergraduate thesis. Our rats were housed one to a cage, each cage fitting a slot in a giant steel grid, where it could be slid in and out with ease. Only one wall of the cage was made of mesh, the wall facing outward into the room. So these guys couldn't see their neighbours. Maybe they could communicate with them through squeals and scratchings. But they must have been lonely. Maybe the proximity of their neighbours, the sounds they uttered behind the steel walls of their cages, made them that much lonelier. I didn't know. They didn't talk about it.

Beth had trained me as well as any rat. I went in. I hung up my coat. I unlocked the door to the inner sanctum and made my way to the cages. My rats were all there, busy doing nothing, as usual. Scratching and whispering, scurrying, hiding, perhaps talking to each other in little rat voices. They paid me little attention. I was a familiar sight, or more likely a familiar odour, and we'd have time enough to visit as the night wore on.

"Hi, little guys. Who wants to go first tonight?" I pulled a cage out from the middle of the grid, just to make life interesting, and carried it to the procedure room, whispering all the way in my rat voice. Then I gently grasped the fellow by the scruff of the neck and held him for a moment beside the chamber where he would perform his nightly duties.

"Hungry, huh? Well, there's plenty to eat, but you're going to have to work for it." I dumped my rodent companion onto the platform that recorded his weight. I filled the pellet tray. I filled the water bottle. I made sure everything was perfect. On a fresh data sheet I recorded the date, February 12, 1977, the subject number, and his weight—before supper. Then I picked up the rat and placed him in the left wing of the experimental chamber, where he regarded the lever, the lights, and the food dispenser on the other side, still out of reach. Finally, I lifted the slide between the two sections of the box and watched, horrified and amazed, that this little rat obeyed so perfectly the commands issued by his brain and his stomach. He did what he was programmed to do. Flawlessly.

I went through a dozen more animals, and I was still only half done. I wouldn't arrive home until nearly midnight. Another long, lonely, boring night. And it was particularly lonely because Sharon and I had been fighting again. Always fighting. My long hours of work made her feel abandoned, like she didn't count. Despite my reassurances—"Of course you count!"—it was never enough.

"But I have no choice!" I'd argue back. "If I don't finish the research, I don't get a grade, or I get an F. And then I don't get into graduate school . . ." When things got difficult, as they had again now, I pleaded for her understanding, for her strength, or if those weren't forthcoming, I pleaded for her to lay off. I didn't want to feel that I was recalcitrant, naughty, unkind, unfair. I wanted her to put her arms around me when I got back, even if I got back at 2 a.m. No more fighting. But now, as I shuttled about the lab, the angry, wounded wrinkle of her brow floated above me, behind me, and the resonance of her nasal voice rose from the hum of the fridge.

The old lab fridge. Sitting in the corner of the procedure room.

Would I? Should I?

No! Once was enough. Somebody would find out. No they wouldn't. The stuff is brown, for God's sake. Nobody is saving it up for the rats, that's for sure. It's going bad. It's probably five years old. Yeah, but it works. It still works. Oh, does it ever. Yeah, and it's probably toxic. You're probably going to die. If you do what you're thinking of doing. Don't even think about it.

But I *am* thinking about it. I can't stop thinking about it. And there were no ill effects last time . . .

The bell went off and brought me back to reality. If this was reality. My first reaction was a rush of shame: it was vile. Shooting some undefined liquid into my veins. Okay, it was morphine. Morphine, the wonder drug. Morphine, the perfect narcotic. The pure essence of which everything else—even heroin—is a derivative. But it was disgusting to shoot that stale stuff in the fridge. A familiar glare from somewhere inside. I picked up my now well-fed and well-exercised little beast, and it seemed as though he was smiling at me: *I know what you're thinking.* No you don't! I weighed him again, a bit more roughly this time, then put him back in his cage. You don't know what I'm thinking, you dumb rat. It's not your morphine anyway.

To get my mind off the fridge, off Sharon, I put the next rat into the box and picked up my novel, plunked myself down on the musty sofa and started to read. Nobody was around. Not only the lab but the whole subbasement was deserted. No sound. Except for the scurrying of those rats still awaiting their moment of glory. And the others, the sated ones, licking their fur contentedly. A sound that grew louder in my imagination: soft tongues scratching and scraping as they cleaned their soft white fur. They were at peace. Like I would be if I . . . No no. Don't go there. Not again.

I'm a big boy. I'm studying to be a psychologist. But I still like to read horror novels sometimes. Especially lately. And Anne Rice evokes the most compelling images. A newcomer has entered the parlour. One of the older vampires crosses the room so swiftly his movements are invisible. He grasps the visitor fondly by the lapels. He whispers to him, part seduction, part warning: "So you want to become one of us? But are you strong enough to bear the curse of isolation that will be yours forever? With a taste of my blood?" And I'm thinking about the morphine in the fridge again, because it is like the vampire's blood: dirty, poisonous, yet offering me its singular powers. It will plunge me into the land that is inhabited by the few, the outcasts, those who prowl by night and sleep by day, whose business is the sating of a shameful hunger. And now other images are awakened. My memories of the Heap, both fond and repugnant. Ralph putting Jim to sleep with a shot of Seconal, a drug that would one day kill him. And my childish wish to be one of them, despite the foreshadowing of destruction that hovered there.

Only fifteen or so rats to go. I'll never make it. Too long. Too tempting.

Don't think about it. Don't think about the little bottles in the fridge. You might never have noticed them if you hadn't been searching for a can of pop. And don't think about the syringes lying so neatly in their paper wrappers in the cupboard. Don't think about them! But I look at the vein in my arm, so rapidly I can't stop myself. Up until a week ago, I hadn't shot drugs for over two years. That's all over. A youthful folly. With its share of horrors, to be sure. Nobody knows the trouble I've seen. Nobody knows . . . but Jesus. I'm actually humming this as I get up to replace one rat with the next. I'm humming this and I'm smiling a little to myself, smiling with a sneaky little smile, a sneaky little rat smile. A smile for no one. A smile no one can see. But there is a quickening in my pulse. A part of me has given up.

———

Addicts spend a lot of time giving up, but not nearly as much time as they spend trying not to give up, trying to hold on. I've been trying to hold on for a couple of hours now, and my brain is losing the battle. Why is it so hard? Why is failure starting to seem inevitable? As we've seen, the ventral regions—especially the OFC and ventral striatum—are where craving builds on value and dopamine energizes neuronal neighbourhoods that shine with the anticipation of pleasure and relief. But the dorsal areas of the prefrontal cortex also play a crucial role in addiction, because these are the areas that keep anticipation in check. The human brain's most respected achievements—judgment and choice, planning and self-control—rely on the activation of an arch of tissues right in the middle of the prefrontal cortex: the anterior cingulate cortex (ACC). The dorsal anterior cingulate cortex, or dACC for short, is in charge of choice, self-monitoring, and the resolution of conflicting goals. The dACC is where context and judgment come together to create the will, that beam of self-direction that makes it possible to choose consciously and act morally. The dACC activates its neighbours to create the fulcrum of decision making, the keen blade of intelligence itself. Should I turn right or should I turn left? Should I fight or should I flee? Should I go ahead and make myself feel good, or should I turn the other cheek and give it a pass, so that I can feel better in the morning? Should I or shouldn't I? This kind of decision making is directly connected to other dorsal regions: the dACC chooses what to do from the possibilities held in working memory, housed just down the hall and around the corner, in the dorsal-lateral executive suites (see Figure 5). The elements in working memory are what I have to choose from: Here are the options, now what's the plan?

The choice leveraged by my dACC this night is a noble alternative to the desire orchestrated by the partnership among my orbitofrontal cortex, ventral striatum, and amygdala, that ventral mafia of

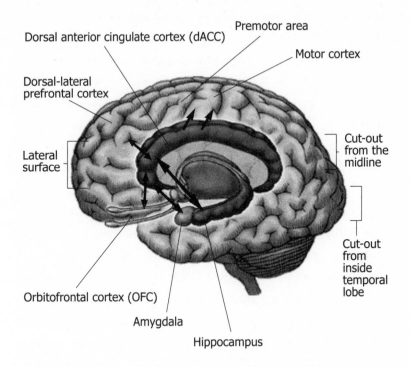

FIGURE 5. The anterior cingulate cortex (ACC) is the front half of the strip of tissue (actually two adjacent strips, one on each side—shaded in the drawing) forming an arch along the cortical midline. The dorsal ACC (dACC) is the upper corner of the arch. Arrows coming and going from this region trace pathways of connection with other structures, including the dorsal-lateral prefrontal cortex and the hippocampus, both involved in working memory.

compulsion. No, don't go for the morphine! Look away, look away. Read your book. Do another rat. Go home, but stay clear of the fridge! My dACC is kept busy with that one. But there's a problem: it's not a choice I make only once. It's a choice I have to keep making again and again, every time I pass the fridge, every time its raucous hum cuts into my reverie, every time the image of Sharon's anger ignites my anxieties. Again and again and again. But so what? Once you make a decision, why would it be so hard to make the same decision again? Any addict will tell you: it just is. But psychologists now have a name for the problem: ego depletion or ego fatigue. In doing its routine chores, the brain uses up a fantastic amount of energy—more than the rest of the body combined. And a lot of that energy comes from glucose, or sugar, which is an important source of glutamate and GABA, those work-horse neurotransmitters that carry messages from neuron to neuron. When the dACC has to keep working to control an impulse, one that keeps recurring, or just won't go away, it uses up its supply of energy. It can't replenish its store of neurotransmitters. *It gets tired.* Very much like a muscle. Try holding your arm out at your side for half an hour. It's pretty easy for the first five minutes, but it gets harder and harder after that. Just a simple physical action, maintained too long, soon exhausts the resources that made it possible.

Ego depletion has been studied for over ten years, and the experiments that psychologists use to study it cut right to the heart of addiction. In one of the first studies, participants were asked to come to the lab hungry. Half of them were allowed free access to a bowl of chocolate-chip cookies but told to ignore a bowl of radishes. The other half were allowed to eat radishes, but not the chocolate-chip cookies. Those who had to resist temptation and avoid the cookies couldn't think as well after an hour: they did more poorly on a puzzle task and they gave up more easily. They had used up their dorsal resources. Their dACC was exhausted by the simple act of saying "No" to temptation. In another study, some participants were asked to control their emotions when watching a movie, while others were allowed to laugh or cry, to express whatever they felt. Those who had to control themselves again failed more easily in a puzzle task right afterward. Controlling emotions, like controlling impulses, tired out the dACC, used up its chemical reserves. And those who lost this capacity were less able to continue to control themselves, to think, to judge, to supervise their own behaviour.

The loss of self-control has long been a trademark of addiction. Now scientists have identified the weak link in the neural chain that makes it so. Brains lose control. People lose control. It sounds like another mind-brain parallel. But just like the last one—the loss of flexibility as synapses and habits rigidify—there's a deeper lesson here: the rules of neural conduct, the physical limitations of brain matter itself, provide the cause. The breakdown of human functioning is the consequence. Once again we find that addiction arises from vulnerabilities in the nervous system. And by understanding the limitations of the dACC, we can see why addicts lose so much more than their sobriety. They lose the mental muscle tone for self-direction, for resolve, for strength of character, and for decency itself.

The dACC is smart. The dACC is wise. My dACC is capable of looking past the attractions of the moment and judging the big picture. It wants to—*I* want to—get home safely, with or without Sharon's blessing. My dACC is my captain, my director, my will. But it's listing, it's sinking, it's going down. Each time the impulse comes up, I can feel the knot slipping. Each time I catch a whiff of the chocolate-chip cookies, it gets harder to choose radishes. After an hour and a half of suppressing my feelings, watching my own little movie here in the rat lab, my control is now flaccid and flawed. I can't suppress my feelings much longer. And I can't solve this puzzle anymore. It's too tough.

My body arches slightly forward as I come back from the cages. I am eager. I am dropping from the stratosphere of the Young Psychologist, the man with the keys, to the swamplands of unfettered craving. And I feel like it's going to be okay, because it has to be okay, because I can't stop it anymore. And maybe I don't give a shit. Maybe I'll just do what I want. There is a thrill to giving up, a singular joy to giving in, a crass satisfaction to changing state from effortful self-control to *what the fuck*. Excitement rises as my ventral striatum lurches into action. The goal *is* attainable. Dopamine surges. It's all yours. To take. Nobody to stop you. Only the rats are here to whisper their disapproval as my hand reaches to the fridge door.

There. It's open. The little bottles of liquid morphine sulphate sit in their shelf, exactly as I left them. Next to the butter, for God's sake! Leftovers from some former experiment. Nobody needs them now. Nobody but me. And it's really problematic that my OFC cares only for this. That there's nothing else of value at this moment. That narrowing, that cancelling of the worth of other pursuits, leaves my dACC to fight alone, without its usual allies: meaning, purpose, interest in the world. The strain is too great. All that focused desire, pitched against faltering self-control. Too much to ask. I pick up one bottle and examine it, and I can see that it's going bad. It *is* turning brown. Is that . . . ? Yes, there's no doubt: it's got a slightly *smoky* texture. That can't be good. I pick up another. Same thing. And I'm going to inject it into my bloodstream? One last reach for dorsal control, like a hand grasping for a rafter. But I can't hold on, and I fall.

I fish the next rat out of his cage, get him going, and then open the fridge once more, absent-mindedly counting the bottles, estimating how many more times . . . My mind is whirring quickly. Fragments of images from '73 and '74 in Berkeley, the junkies I met after returning from Calcutta the first time. My friend Eddie. As if a junkie could ever be a friend. All big eyes. Big eyes and lies. But he and I became soul brothers for that drive to downtown Oakland, where his suppliers were hidden away, in bricked-up slum housing or some place nearby. I never knew for sure because no white boy was allowed up those stairs. I waited in the car. Eddie went up, and he took much too long. When he came back I knew he'd already cheated me: taken some of the smack for a "taste" — just to make sure it was, you know, like, real. The big eyes with little pupils, barely visible against his irises. What had he added back in to make up the difference? And even then I didn't care. I was that eager. But there was one time, one score, when I was allowed up. There were four big black guys standing around a table: serious junkies, shooting fifty to a hundred dollars' worth at a time. I was scared. They might just turn around and knife me. But I stayed in the background until I was summoned. "Hey man, you want the cotton? Thas what's left." I got whatever liquid was left over in the piece of cotton still lying in the spoon, after another guy had sucked eighty dollars' worth of smack through it. A guy with shiny yellow eyes, eyes that screamed "hepatitis." And even then I didn't care enough to stop. Even with the taint of disease simmering in the spoon, rising into the needle.

The phone jars me. It's Sharon—who else? The dream fades with impossible speed, the bubble image of sublime engulfment now shrinking to nothing as it recedes to the horizon of my thoughts.

"Are you done yet?"

"No, I'm not done," I reply quickly, maybe sounding manic, edgy. "I've still got over a third of them to go."

"Why is it taking so long?"

"Sharon, it always takes this long."

"No it doesn't. If you'd started at seven like you said you would, you'd be done by now. What time *did* you start?" That old, familiar anger/shame cocktail now pushing up through my chest. I want to yell at her: *Leave me alone!*

"I started a little late."

"How late?"

"I don't know. I didn't notice."

"You know how this makes me feel—" and now she's whining and I'm stammering.

"Yes, but . . ."

"Why don't we meet for coffee? I'll come down and meet you at the donut shop."

"No, Share, I don't want to . . . That will put me back an hour. I just want to finish and get home."

But I'm not stammering now. Why so brave all of a sudden? Little bottles, in the fridge. Is that it? The part of me that's going to win, any way it can, approaches like a huge, engulfing shadow. It's going to be all right, isn't it? No reason to get upset when you can have it all.

Sharon is going on and on: "Why aren't you ever there when I need you? Why can't you just do this little thing to help make me feel better?"

"Look, I'll be home as soon as I can," I say, now with some weird calm. Because, while I definitely don't want to meet for a coffee, while it's the last thing I want, it's not just because it will delay my work. It's not just because I've got nothing to say to her, nothing that hasn't been said a thousand times already. It's because I've given up. The struggle is over. Dopamine now rising up in me like sap, past the threshold, past any chance of stopping. It's going to happen. It's going to be all right. And then I'll meet her. On my terms.

I had to put the phone down when the rat buzzer went off. I had to cut her off, mid-plea, and tell her that science demanded my presence. A bit too lighthearted for the occasion, but I hardly cared now. Everything was starting to transform. I wasn't lonely anymore. I wasn't upset, or mad, or anxious, or even ashamed. And that was the most surprising thing. My shame was gone. Released, not by taking the drug, but by abandoning self-control. As if self-control were the critical parent, my mother's judgmental gaze, ever-present . . . until now.

I was free.

I held my rat tenderly and walked to the scale, placed him in his compartment, adjusted the counterweights, listened intently. Nobody around. Not now, not ever. And I replaced him in his cage. I veered away from the cages. The next rat would have to wait, and the quickening was everywhere, in my blood, in my throat, in my breathing, fast and shallow. I could feel myself start to sweat as the moment approached, an orgasmic joining of the two waves, one from the shore, one from the sea—one of glutamate, carrying meaning from my OFC, one of dopamine, carrying thrust from my VTA—waves rising higher and higher before they would combine in my ventral striatum . . . and fall to pieces, the swimmers screaming in delight and fear.

I peeled the paper wrapping off a syringe. It would hold three millilitres, plenty. I attached a needle, also peeled from its wrapper. It was the perfect size. It would not hurt. I shook the bottle and looked away. Were those . . . particles . . . in there? I stuck the needle point into the rubber bull's-eye in the cap of the bottle, turned it upside down, and pulled liquid into the syringe. Enough. I squirted the liquid into a tablespoon, added a bit of cotton wool, and pulled it back into the syringe through this homemade filter. I didn't bother with a rope or belt. I needed to do this fast, before I could think too much. I could pump up my vein just fine by clenching the muscles in my fist and forearm. The liquid went into me. Easily. I watched almost from outside. Almost detached. In fascination. What am I doing? And then it hit.

When you shoot morphine, two things happen at once: your scalp starts to tingle and itch like crazy and your legs turn to lead. Next, everything goes soft. This was okay. This was good. No ill effects. The only effects were . . . lovely.

For a little over a year, I remained psychologically addicted to morphine. Once the bottles in the fridge were gone, I found two large jars of morphine sulphate powder in another room in the animal wing. I couldn't believe my luck the first time I found them. It was an obscene amount of the drug. Big fat jars, peanut butter jars, filled almost to the top. The first few times I took a little out, I told myself it would be the last time. But "the last time" kept skipping to the next time and then the time after that. And then I stopped telling myself anything. I just went with the flow: craving, relief, boredom, islands of self-control, then the same cycle once again. I never took it for more than a few days at a time, so I never got much of a physical dependency. What I got was a building tide of need, desire, craving — call it whatever you like — that could only be resolved by taking more.

I didn't always shoot it. Sometimes I snorted it. Sometimes I drank it. But shooting it was the most pleasurable. It provided the most dopamine by far: the meteoric drop down the first big hill of the roller coaster and, more and more predictably, the exquisite excitement that crescendoed with the ride up that hill and the frozen moment at the top. I became a maestro of anticipation and execution, devising with increasing skill the lurid scenarios that made it more interesting. The context, the stage props, and especially the duplicity, the rich mix of legitimacy and violation. I relished these scenarios because they were my way of saying *fuck you* to everyone else in the world. I would shoot morphine in the university library, sitting on the toilet. Then I would walk through the stacks, intent on the subterfuge. Devoted student or despicable addict? Who would ever know? Who could even guess? Or in the shower. "Coming, dear! Almost done!" Sometimes I told Sharon about it and sometimes it remained my dirty little secret. I always wanted to tell her, but I couldn't always face the barrage of rebuke that would follow. And though I admitted to Sharon — it was obvious, after all — that I was the sick one, the one who was blameworthy, I couldn't be sure what was causing what. Was our difficult marriage fuelling my addiction, or was my addiction destroying our marriage? It eventually dawned on me that both were true.

Rats will often consume morphine and other opiates, but until a famous experiment in the late seventies, most psychologists never thought to ask why. While some rats were housed in the usual sort of cage, made of cold steel wire, and isolated, one rat to a cage, others were housed in a "rat park": a large open-topped wooden box full of wood shavings and other diversions and, most importantly, together in a large group. All rats were offered water or morphine to drink. The choice was theirs. But the rats living in Rat Park drank a lot less morphine than their counterparts living in isolation. Presumably because they had fun and they had company in their daily lives. The authors saw addiction as a product of an impoverished environment, not a characteristic of a particular class of drugs. They felt that companionship made life pleasant enough that external opiates weren't needed. And maybe they were right.

My long nights at the lab were pretty lonely. But what made my life most miserable wasn't isolation; it was the ongoing arguments with Sharon and the shame and anxiety that lingered in their afterburn. Social attachment, whether between mothers and babies, lovers, friends, or litter-mates, is fuelled by internal opioids: its measure is the level of relaxation and warmth provided by those tiny molecules. And a key buttress of attachment is a good feeling about yourself—a "positive self-image," as psychologists like to call it. If you feel good about yourself, if others seem to think well of you, the fabric of social warmth is sure to endure. But when you feel inadequate, when nothing you do seems to please those you care about, internal opioids plummet. I didn't feel good or adequate. I didn't know how to find the warm, connected feeling, except by submitting to Sharon's demands. And another part of me recoiled at that. My internal opioids were in short supply. In a sense, I was in a state of withdrawal regardless of my drug use. And the only way I could find to fix it was with another supply of opioids, taken from the lab shelf. I can make myself feel warm, no matter what you say or do. Except that stealing drugs and shooting them in the shower didn't leave me feeling particularly virtuous. When those drugs left my system, the bitter sting of contemptibility came flooding back. I was a liar and a thief, and that wasn't going to win me attachment points with Sharon or anyone else.

14

CRIME AND PUNISHMENT

By April 1978, my life was hurtling down two tracks stretching farther and farther apart. I had been accepted into graduate school in clinical psychology at the University of Windsor, a four-hour train ride from Toronto. My dream of becoming a psychologist was coming to fruition. At the same time, the enmeshment of my drug taking and my marital problems made life unbearable for both Sharon and me. I knew beyond any doubt that I had to resolve that enmeshment if I wasn't to destroy myself and wreck the future I was starting to taste. I decided that I needed a break. A separation—not a permanent separation, just a few weeks on my own. I wanted to get off drugs, and that just didn't seem to work with our endless cycle of fighting. Not that I'd been tremendously successful prior to life with Sharon. But maybe now would be different. In any case, drugs or no drugs, I needed to get away from her. To walk, to meditate, to study. To find out who I was or who I could be when I wasn't on the defensive. In fact, that seemed the only hope for our marriage as well.

But nothing worked. I flew to Boston for a month off, with Sharon's consent. This would be our "growing period." In five days she called, desperate, begging me to come home. An hour and a half of intense need poured out of the phone. She cried, she sobbed, she told me how alone she felt. And she didn't miss any opportunity to blame me for it.

"All year I looked forward to spending the summer with you. And now that summer's here, I'm alone," she wailed.

"But it's only for a month."

"It's just so totally selfish of you to want to do exactly as you please. What about *my* pain?"

"But we agreed to this . . ."

"Okay, I was wrong. I can't take the pain of feeling you're doing nothing for the relationship when I have to sit here alone, waiting out your whim."

I held the phone away from my ear and examined my feelings. Sharon seemed almost unfamiliar to me. This wasn't the person I loved. My pity for her seemed like an emotional habit, but I felt little actual concern. I wondered how much love was still in me. And I began to realize that I was really trapped. I didn't just feel trapped—I *was* trapped. Crushed by the collapsed coal mine of her needs. There was not going to be a happy ending after all.

I returned to Toronto and things got worse. We fought several times a week. The more Sharon demanded that I prove my love, the less love I could muster. I ricocheted among taciturn withdrawal, bright anger, and guilt, always guilt. But then, at times, the air cleared and I found my affection for her. I still wanted to make it work. It still seemed possible. If only the burden of her pain were not entirely on me. If only I could find my will and stand my ground: This is how much I can give you, this is more than I've got to offer.

We started seeing a marriage counsellor. Leonard was good. Leonard of the long sideburns and perfectly brushed-straight-back grey-black hair. But Leonard was no magician. I kept taking drugs. Not every night, but often enough. I had spent the year doing volunteer work at a local psychiatric hospital while finishing my psychology courses. I had been entrusted with another experiment, another set of keys, and another opportunity to roam the halls at night. As my morphine supplies dwindled, I became more creative. My master key opened office after office, and I availed myself of prescription pads. It was not terribly difficult to forge a prescription. I copied the symbols from one I'd gotten from a doctor, and I succeeded in getting various things: hydrocodone cough medicine—an opiate, not a dissociative—and my old friend Percodan.

It was nerve-racking, walking into a drugstore with a fake name and a fake address. One time, I gave my name as Joseph Tesher (someone nonexistent) and made up an address on Bathurst Street. The pharmacist was friendly. He chatted while pouring a large bottle of Tussionex, a strong narcotic cough syrup, into the four-ounce bottle that would soon be mine. He typed the address dutifully onto the label and then slowed down, jerkily, like a car in the wrong gear, lurching in little fits and finally stopping.

"That's the . . . that's the . . . that's the apartment on the west side?" he asked. "But there's no such number on Bathurst Street. No such building."

"Sure there is." My voice remained calm but my heart was suddenly pounding.

"No . . . ah . . . no, actually, there isn't. I live . . . I live right near there, and . . ."

"Well maybe I got the number wrong. You see, I'm just staying there for now. Just visiting. With my grandparents. Maybe it's—"

"Maybe you mean 2100 Bathurst Street," he offered. He wanted it to be okay. He really did. I felt badly for him, but I couldn't stop now.

"Yes, that's it! Sorry about that." He continued his work, but he didn't look up at me again. I have no idea why he didn't just stop and kick me out. Or call the cops. I wasn't the least bit threatening. But he seemed far more frightened than I. He muttered under his breath,

"T for Tesher, T for Tussionex . . . T for Tesher, T for Tussionex," a mindless chant. I was afraid that he was going to have a breakdown right then and there.

Through these and other adventures, I'd found ways to get drugs when I wanted them badly enough. And now, trying to surf the ever-stormy waters with Sharon, the urge, the means, and the opportunity buzzed around each other, sometimes connecting long enough to yield another outing. Or evaporating, leaving me empty-handed, disappointed, depressed. On those nights I might get a bottle of codeine-Aspirin tablets, sold without a prescription. I found a way to filter out the Aspirin and grant myself the respite of a codeine high. Sometimes Sharon would join me. Nobody is immune.

Sharon and I were preparing to move to Windsor. I was going to start graduate school while she began her job as a social worker. The planned move seemed a last chance to bind our lives together and make the marriage work.

But it would not work. It refused to work. The chemistry was all wrong. At one point I got so sick of the whole thing, and so frustrated with my failed attempts to get off drugs, that I decided to enroll in a six-week inpatient recovery program. Sharon wouldn't hear of it. To her, this was just another vacation, another excuse to ignore her needs and focus on myself instead.

"I'm just telling you," she said, "I won't be there when you get out."

"Good!" I spat back. "You don't care about what's inside of me, as long as I'm always around." My confusion mushroomed. Didn't she want me to stop? I balked at her threat. I wasn't quite ready to give up the marriage. Yet some of my anger must have turned to hatred.

———

On a warm night in August, our fighting escalated to full-scale war. How had it gotten so bad? I couldn't stand it anymore. I told Sharon I had to go . . . somewhere, anywhere.

"No, you can't just leave like that!"

"Just for the night!"

But I soon realized I wasn't going to get permission. I had to act unilaterally. I left her in her mother's house, in a fine residential district in Toronto, sobbing at the kitchen table. Nobody else was home, but I told myself she'd survive. I didn't look back. I didn't think of packing. I got my wallet and my keys and went out to my car, an old VW squareback, parked just down the street.

That's when things began to fall apart. I started the engine after a couple of tries, but before I could pull out, Sharon opened the passenger door and got in beside me. She was enraged. She sat there, inches from me, spewing the most insulting language I had ever heard from her. I was a coward! A selfish asshole! A gutless drug addict! That's how you handle your problems. Not by talking. Not by compromising. But by drugs! When I'd absorbed enough, I ordered her out of the car. I was yelling by then, too. She finally got out . . . and lay down in front of the wheels. I couldn't believe it. I got out and saw that she was lying on the street, just below the front bumper. I begged her to get up. My voice sounded staged, echoing in the silent street. I said this was ridiculous. I could not back up. There was a car parked right behind me. And I couldn't go forward without killing her. Finally she got up and jumped back in the passenger seat to begin her tirade once again. She said she would not allow me to leave. She lay down in front of the car a second time and got back in to berate me a third time. Why hadn't I locked the door?

I finally lost it. For the first and only time in my life, I struck a woman. It happened so fast that I was not aware I had done it until I registered the shock on her face and the sting in the back of my hand. The words stopped. There was nothing more to say. She held her face, the skin around her eye already starting to swell. I rushed her back into the house and we did the normal things people do when someone gets hurt—the washcloths soaked in cold water, the bag of frozen peas. And we called our marriage counsellor. Emergency session, Leonard! You have to come . . .

I was more shocked than Sharon. But my anger had vanished. Within a few minutes I felt little of anything. No real pity, some shallow semblance of regret, circling confusion. I knew that something was very wrong but I wasn't sure what. It must have been close to midnight, but Leonard said he would be there in forty-five minutes. I became restless, unable to sit still. I wanted drugs. I apologized for the tenth or twentieth time and then told Sharon I had to go for a walk. I had to clear my head. She didn't seem to mind. She knew I'd be back in time for Leonard.

I pushed open the door and felt the warm air embrace me, carry me like a child away from this place. The breeze was alive with some familiar essence, an invisible presence, whispering. Like the nights at Tabor when I had wandered through the dark streets. Like the night I'd gone into the empty church and sat there in the dark, feeling unreal. I walked down the block, turned to my left, walked several more blocks. The streets were completely empty. I lost all sense of myself, as if I were simply a character in a play.

Where did it come from, that first time? The impulse. The plan. Surely my dACC was completely shut down through the ravages of ego fatigue. The circuits of self-control could not have been operating. I had tried too hard for too long to control too many things: feelings, impulses, actions. The dACC is connected to many other regions of cortex, and directly to the limbic modules of motivation and memory: the amygdala, with its emotional associations, and the hippocampus, with its episodic storage sites (see Figure 5). From the hippocampus it retrieves alternatives, and its connections with the dorsal-lateral prefrontal cortex edit those alternatives and hold them in working memory. Its tools of selection and choice can then operate on a limited pool of options, and it can use those tools to control the emotional uprising centred in the amygdala and OFC. Meanwhile, its connections up through the premotor areas to the motor cortex turn intentions into actions. The intention *not* to do something translates into a braking manoeuvre: the muscles clench and hold tight; the hand, whether balled in a fist or open to pat, comes back to the side; the tongue freezes mid-insult. That's how the intrinsic intelligence of the dACC achieves its control over the outflow of acts. And it's no accident that the same mechanism is the requisite for morality. Most of the Ten Commandments start with something to the tune of "Thou shalt not . . ."

But it wasn't working for me. Not tonight. Ego depletion had passed some irreversible threshold, and I'd lost my capacity to *withhold*. The tissues of my dACC were depleted of neurotransmitters. My limbic system was operating on its own, without guidance. This was one version of the state we've called dissociation. Dissociative drugs serve to dissolve the bridges between the cortex and limbic system, and the dACC is one of those bridges. That's how the dACC imposes sense on meaning to achieve its function of impulse control. But tonight I managed to shut down the bridge without drugs. I was in a dream. I'd gone limbic.

I saw a dark house, a large dark house, that reminded me of something from my childhood, and beside it was a kind of TV antenna, popular in those days, with triangular rungs that you could climb as easily as any ladder. I climbed it without thinking. The only emotion I felt was a kind of release. I climbed to the second floor, and I saw that the window was open a crack. I pulled it up the rest of the way and went inside. I found myself at the end of a dark hallway, empty except for the faint odour of simmering meat from hours before. A sweet smell. I walked down the corridor, listening for any sound. There was nothing. I was inside this house that wasn't mine, I didn't even know if it was occupied, and I walked toward the bathroom, to the medicine cabinet. Nobody was there to stop me.

There was a door leading from the bathroom to a bedroom, and I thought I heard the soft, regular breathing of someone sleeping. I closed the door soundlessly and waited till my eyes adjusted to the light coming in from the street. I opened the cabinet softly and saw the bottle of Percodan clearly on a shelf. I wasn't surprised to find it, because this seemed a natural extension of the dream I was in. I put it in my pocket and tiptoed downstairs. I let myself out the front door and closed it gently behind me. That was all.

With a few Percs in me, I was able to coast through a two-hour session with Leonard and Sharon that night, peacefully, gracefully, willing to accommodate wherever I could. I did not feel remorse. In fact, I felt triumphant. My prize was inside my blood and brain, where no one could find it and take it away. We recited formulas and listed rules, modifications of past attempts, we updated contracts and commitments, and then Leonard was gone and we were alone. I couldn't believe either of the crimes I'd committed that night, but I was not upset. It wasn't just the drugs. I had really given up control, and so there was nothing more to regret.

Sharon and I moved to Windsor in early September. By mid-October I decided I had to leave, at least for a few months, and I rented an apartment within two weeks. Now it's mid-November and I still haven't moved out. What is this morass, this sticky substance pulling me back? Where is my energy, my motivation? I cannot seem to find the incentive to buy paint. The place needs painting. Some drunken asshole scrawled a swastika on the door-frame. Nice. That's why it's so cheap. The last tenants were members of some local gang. Great. What if they come back? No, it's safer here, with this woman, whom I love, whom I lie to, whom I hate. I know I have to leave. It's only a matter of time. She's trying to delay it of course. Every trick in the book at her disposal. But right now I don't care what she does. She's . . . she's the devil I know.

I am still in bed and it's noon. I have a class at two o'clock. And I still haven't decided if I'm going to try to make it. Images from last night flare in the storm clouds behind my eyes. Last night. My God, what did I do? I went to a low bungalow of an office building — all doctors' offices — but almost shabby. Made of plywood. All the better, I thought. And one of the outside doors was unlocked! So I was in this hallway and it was at least 11 p.m. Nobody coming around tonight, that's for sure. But how do I get into the offices? It's remarkable, really, that these doors are so flimsy and they're held by these stupid little locks, not even a deadbolt. So if I were to really throw my weight against it . . . *Ow!* Bad idea. Shoulder sizzling with pain. But the hallway is so narrow that, maybe, if I lie down on the floor and put my shoulders against the opposite wall and really push with my feet, the door will . . . open? Crazy fool. Holy shit. It worked. The door flies inward on its hinges. It doesn't even look damaged, and that's good because . . . I'm inside the first office. I put a desk lamp on the floor, so as not to attract attention, and turn it on. Only for a couple of minutes. It will be okay. But there are no drugs here at all, except for some lousy tranqs. Fuck! What is this guy, a psychiatrist or something? But now I know the trick. Back to the hall, looking up the specialists on the board by the front door. Family doc, dentist, dermatologist . . . Hmm, I wonder . . . ah! "Physician and Surgeon" — oh yeah, I like the sound of that. Sounds like Demerol to me. The door gives just like the last one and I'm inside, but I think I hear a noise. *Oh shit.* A car slowing down outside. Back to the hallway, fast. I'm not going to give up yet. Even if I *could* get out without being seen. I make sure both doors look undisturbed, then I'm back in the surgeon's office. I close the door and lock it. It looks normal, except for a few splinters, but you'd have to look closely. Shit! Lights at the window. If it's the cops, I can't leave, so I get down under the office desk and watch flashlights crisscross the windows slowly and then they're . . . gone? I sit there, completely in knots for fifteen minutes. But there's been no sound since the car drove off. Still nothing. That's it. They're not coming back. Or maybe they are, with reinforcements. So my chances of getting busted are—What? One in three? No, they're not going to check every office . . . I wait as long as I can. Gradually, peace returns. More than peace. Because I'm inside this dark warm place and I know there are drugs here. I can feel it. Someone's office, someone who tried to keep all the drugs to himself. Image of a thick, fatherly figure (not my father) somewhere, hovering. But they're mine now. All mine. Liquid warmth settles, a soup of images, dark and familiar. I almost feel his blessings. That's okay, son. I understand. Like hell he understands. He's going to be *so mad.* I am a little boy rummaging, drawer after drawer. And there are drugs here. So many. Sure enough, drawers full of boxes, piled high, free samples. Must be. And ohhhh, there's the Demerol. Multidose glass vial: 50 milligrams per millilitre! That's the strong stuff. Almost full. Now, the apparatus. Drawer full of syringes and needles, each cozy in its wrapper. I am literally chuckling with glee. I am pretending to be Mr. Hyde, or I'm not pretending. You're *fucked,* I tell myself. But I'm still smiling. The accusatory voice has no power now. No mother, no father, anywhere. And look, a nice folded plastic bag. I start to stuff it. Halloween in Drugland. My mood is off the charts. Intense excitement, glee, power, triumph, and anticipation of the . . . oh yeah . . . shooting Demerol is just so nice. There is nothing like it. I once read that if there's anything nicer in the universe, God saved it for himself. And look—what's this? Dexedrine. Sure, why not? That will mix very very well. He's even got Percs. What does he do with this stuff? Into the bag. Close all the drawers. Getting ready to leave. Putting everything back the way it was. Nice and neat. Isn't that odd? I'm stealing his drugs but I don't want to leave a mess. Don't want to *disturb*. But of course he's going to notice that the Demerol is gone. Probably tomorrow. Oh well. He can always get more.

Home again, home again. But which home? My apartment, yeah, my apartment—that's a laugh—I open the door with my key. I avert my eyes from the swastika. It's freezing in here. So what? I wipe the kitchen table with damp paper towels. Put down my bag. I swallow three tabs of Dexedrine (milder than methamphetamine, but with similar effects). I load up the syringe with three millilitres of Demerol. Humming. Flying low. Dopamine buzz because the goal is getting closer by the moment and the image fragments are motes of dust, coalescing clouds in my consciousness, and my ventral striatum is lifted between the waves of glutamate and dopamine, meaning and motive. Norepinephrine way up: I'm crazy with excitement. Not just anticipation, but a delicious blend of anxiety and triumph, the icing on the cake. I take off my coat, pump up my arm, prick the vein right on the sore red spot left from a few days ago, suck up a little cloud of blood, all systems go. Ready: push the plunger steadily all the way. Before it touches the bottom of its trough the waves of warm, thick pleasure start their massage. And now, a moment later, they are no longer gentle. They flood me. *Ahhhhhhhh.* So good. Demerol—completely synthetic, marvel of science—hits my opioid receptors and pumps them full. Loaded, weighted leaden liquid colourless except for the whispering of dreams in this half-gone state, almost as severe as the cut-off of consciousness from nitrous oxide but so much richer and fuller, softly crushing density at the bottom of the ocean, and it keeps getting deeper, wave after wave.

For three hours I roam my soon-to-be apartment, shooting more Demerol every half-hour or so. Craving builds so fast, just as the tide begins to recede, leaving room for another slam-dunk, until I'm barely conscious. I'm in a stupor. I can't go home to Sharon like this. A few more Dexies and my pulse quickens enough to cut through the fog. Can I drive? I hardly care. But the clutch is some devious device meant to befuddle me, so very carefully I let it out, and I'd better just stay in second gear. I drive home in second gear. Idiot. If the cops see me driving this slow they'll surely pull me over.

But I get home in one piece. Sharon's asleep. One more shot nestled on the living-room sofa and I lie back and drift: too much Dexedrine to sleep, too much—way too much—Demerol to do anything else.

And that's where she found me at 4:30 a.m., when she got up to pee or to worry or whatever she does when I'm not around. Found me, dragged me to bed. Scalding words this morning. Like I care. Pillow over my head until she finally leaves for work. In and out of sleep since then. Have to pee. Face in the mirror is not a pretty sight. Not the face I'd like to present at Psychology 1250: Cognitive Processes. I would feel so incredibly shitty right now if . . . if I didn't have that bag of goodies to see me through the day. Just the thought of it brightens me. It's going to be okay.

Those four months—the long, painful wrenching apart of our lives from September through December—were the period of my descent into a spiral of crime that seemed unfathomable. This wasn't me. I wasn't a criminal. And yet it *was* me, and I *was* a criminal. A graduate student in psychology by day and a thief by night. The pattern gradually acquired its own logic—a logic that began to seem inevitable, almost natural. First I broke into someone's house in Toronto—a crime punishable by years in prison—and it weakened some part of me, a part that remained permanently damaged. Then, in some bizarre way, I began to own it, to use it intentionally, tuning and adjusting it by trial and error. That first night—sitting comfortably across from Sharon in her mother's elegant living room, greeting Leonard as he bustled through the front door and froze at the sight of Sharon's black eye—brought together a cluster of emotional gains, fused into a nugget of intense meaning, that worked so well it never fully dissipated after that. Meaning that not only stayed intact, carved into my synapses so deeply by the intensity of events, but that continued to crystallize and cohere with each break-in. I had discovered some ultimate act of defiance, an expression of rebellion, withheld too long, that ended with the satisfaction and relief of an opiate orgy—a disconnection that led to a reconnection. All of the attributes of my drug-seeking till now were embodied in that act. I did not ask. I took. Without the intermediary of anyone's love or approval, I got the warmth I wanted. But there was a new element as well: pure, savage defiance. That was the gold ribbon that tied it all together. My crimes, like those of a young child, brought together the need for warmth, connection, and safety with a roar of protest. You can't treat me like that! You can't keep me away! I'll show you!

I grew up next door to my three girl cousins, and I secretly wanted to be in their family so that I could live with them forever. There was a warmth and richness in that household that called to me. My own house seemed subdued, like a fire almost out. But the Mindens' fire burned brightly throughout the day and night—I could see it in the yellow light filling their windows when I went to bed. I must not have been content to stay on the outside, looking in, because one summer afternoon, when I was four years old, I did something terrible. I picked up a metal bar I'd found in the backyard and proceeded to break their windows while they were out for a drive. Every window I could reach. My aunt and uncle and cousins came home to a carpet of glass splinters covering their basement and first floor. I had hidden the metal bar, so I didn't see how I could be suspected. But of course I *was* suspected. And spanked like never before. Nobody ever understood how I could do a thing like that. Even my mother, who knew me best, couldn't quite fathom my discontent, my impulsiveness, my recklessness, and my urge to destroy the barriers that kept me out of someplace warm.

Defiance turns defeat into victory, but the price tag for me was formidable. It hinged on a state of dissociation that I could retrieve, more and more easily, by throwing a switch, disconnecting my dACC. Which was hardly working anyway. When I wasn't suppressing my appetite for drugs, I was holding back the anger and grief and fear and guilt geysering from the gulf between Sharon and me. That's a lot of suppressing. And my dACC was by now badly damaged. Its function was broken and repaired, broken and repaired, until, like a shelf fixed too many times, it could not bear weight. So I would do it again. But another break-in, another theft, left me more alone, more ashamed, more terrified than before. And so the need for relief continued to escalate, while the task of suppressing it became more and more difficult, until I gave up, put on my coat, grabbed a screwdriver, and raced out to my car.

Yet defiance has its rewards. The singular excitement and sense of power that came with prowling, breaking in, and stealing drugs had all the properties of rushing on methamphetamine. Recall that meth provides a magical mixture of dopamine (focus, power, energy), norepinephrine (the shiny presence of novelty and excitement, refreshed every second), and serotonin (filtering out the overload, relaxing and regulating). This is what I felt. My crimes had the neurochemical composition of a meth-opiate cocktail. And that high, triggered by the rejection of self-control, the disengagement of my dACC, had the instantaneous effect of banishing shame and fear. After struggling to refrain, to stop myself, for hours, pushing those painful emotions back down as they continually rose into my throat, with ego fatigue weighing ever more heavily, with the sense of inevitable surrender splintering the pathetic props I'd erected against it, I would finally give up the struggle. And the shame and fear would vanish! The night air dissolved them.

There was one more benefit to my crimes. Each time I repeated the cycle, I would spend two or three days high as a kite. Then I'd run out, come down, go dry. And I would feel like shit. Hungover, foggy, empty, depressed. Often with the bonus of physical withdrawal symptoms: tearing eyes and hyperstimulation, shortness of breath. My stomach, once it started working again, would have its revenge as well: searing cramps while doubled over on the toilet. But my soul took the biggest hit. The face looking back at me from the mirror was pale, dry, hollow, and aging. The tender skin under my eyes acquiring folds that did not go away. There was one particular emotion carved into that face almost permanently now, one I had never seen there before. It was fear. I had made myself sick: psychologically weak, physically wrecked, without will or warmth or anything resembling self-respect. "What's the benefit in that?" you might well ask. It's that self-destruction paid off some of my guilt. If I'm suffering this much, how can I be responsible for Sharon's pain? And besides, I could reattach myself to Sharon, at least for a while, because now I really needed her, just to hold the pieces together. I could give her what she seemed to want: reciprocal need.

My sins seemed implausible, outlandish, to me more than anyone. Once I woke up on a sofa in the student union lounge. Something was burning. It smelled like rubber and hair—disgusting—and there was smoke coming up from the sofa I was lying on. It took me a long time to realize that my cigarette was still smouldering inside the hole it had burrowed beneath me. I had conked out after shooting too much Demerol, and the faces around me were alarmed, confused, revolted. A friend came to help me paint my new place. I shot Demerol every few hours during his stay. He wasn't my friend after that. I learned how to use plastic to open locked doors. I would cut a piece out of a thin Tupperware container, about twice the size of a credit card, with the ideal strength and flexibility to wedge between a door and its frame and depress the tongue of the lock that held it. Deadbolts and electronic alarms weren't used in most medical centres. I became more deliberate and more skilled. Instead of breaking in at night, leaving defunct doorknobs or splintered wood behind, and imagining the horror and disgust my doctor victims must have felt the next morning, I would enter the building at 5 or 6 p.m., use my Tupperware tool to get into an office already closed for the day, and wait there. By 7 p.m. there was no one left but the cleaning staff. And me. I dressed well. When cleaners or janitors saw me, they thought I was a doctor working late. Even during the act of riffling through drawers. The door suddenly opened mid-riffle one night, and instead of shouting with alarm, my visitor said, "Sorry, Doc. Didn't mean to disturb you! Don't work too late, now. I mean, you're the doctor, but take it easy." This manic-depressive existence went on and on. But I still had hope. Days would go by when Sharon and I didn't fight but instead took pleasure in each other. We watched movies or made love, we talked, we tried. But these bright spells never lasted, and the disappointment was most severe after hope had briefly flared.

It's late December and I still haven't moved out. The cycle has continued and gotten worse, seemingly impossible to stop. Leaden depression, worse after each break-in, crushes me, weakens me. Now I lie here in Sharon's bed, alone. Every time I move my body, even a hand or a leg, dark molasses settles in around the new position, filling the tiny space that has just opened momentarily, obliterating the air, the fragment of peace, that might have entered. There's nothing at all to do. No point in moving. No point in reading. No point in going back to sleep or in being awake, in going outside or staying in. My body is filled with darkness, an enormous blood clot that occupies my interior. Outside: nothing. Inside: thick dead blood. But now my legs are jiggling, and anxiety, like static electricity, arcs through me. How delightful, says a sarcastic voice: a little neon panic to light up the empty streetscape of my psyche.

Last night's fight: Sharon's dark, tortured face twisted in anger. Was it my fault? Was it hers? Who can say? How could it matter? There's nothing left here. There is no marriage, no bond, no shared hopes, no place in which to be together. Now she's gone to work and the house rings with emptiness. I wait until the door slams before I move from bed. I've ignored the discomfort in my bladder until I'm sure I can get from bed to bathroom without seeing her. The thought of her fills me with such unhappiness, but the thought of being alone is no better.

Which leads, this morning, to one thought: Should I? Shouldn't I? Should I? Shouldn't I? I'm so depressed I can't be blamed. No human could be expected to act differently. If he could, if he knew a way to soothe it, he would try. But four or five days from now, when I run out, when withdrawal symptoms begin again, it will be as bad as this. But what else am I to do? How to bring colour and motion to this day? Except by planning another raid, going through the motions, living for those few bright hours between seasons of darkness.

I finally moved into my apartment in early January. It took many weeks before Sharon got the message that I wasn't coming back. She was miserable, her need immense. I was past guilt. We had tried as much as anyone could. But the clouds began to clear. There was a self here underneath the fog of depression and anxiety. The apartment was freezing. There was one heater—a tiny furnace—in the living room. Its heat had to reach the bedroom on one side and the kitchen on the other. But I was warming up on the inside. My life clarified, simplified, the sediment drifting to the bottom, leaving a space where I could study and read. I meditated in the mornings. And I stayed away from drugs. For about four months I was entirely clean. I drank whisky when I felt sad and lonely, and that seemed so . . . normal. Something I was almost proud of. I sat in my cold living room and made up songs, taking sips from a bottle of rye. I felt whole again, and it was a feeling I cherished.

It wasn't always easy. When cravings or opportunities arose, I had to beat them back down quickly and relentlessly. I saw a psychologist once a week: a kind, twinkling man by the name of Dr. Balance. (That really was his name!) He helped keep me on track. I did reasonably well at school. The term was ending, and I went through the arduous process of nailing down a summer internship. Come June, I would be a psychology intern at Lakehead Psychiatric Hospital, in the city of Thunder Bay, about twenty hours by car from Windsor or Toronto. I saw Sharon now and then, but that horrible tearing was over. She was pulling herself together, too. We could have an occasional lunch together. I began dating a woman named Tina, though both of us feared too much connection. I went out with other women. I went to bars and did the crazy drunken things university students are supposed to do. Finally, the lurking threat of falling back into drugs began to recede.

But at the end of the term, partying at night and hungover by day, I let my defences down a few seconds too long. I was into my third drink at someone's house party. I went to the bathroom to pee, and then, on an impulse, I opened the medicine chest, and sitting there staring at me was a bottle of Tussionex. My old friend. It took about five seconds to pour half of it down my throat. I needed to do it fast, before I could stop myself. And then I was high for the next twenty-four hours, and I liked it. Of course I liked it. I had always liked it. For the next two months, the urge to do more break-ins returned with a vengeance. I let myself slide. Depression returned with its own dark magnetism, more intense after every fall.

I moved back to Toronto for a month and a half. A cornucopia of new medical centres to invade. I flirted with suicide. In one medical office I availed myself of liquid cocaine hydrochloride. I shot it. I kept pressing the plunger long after the rush began accelerating. I kept pumping it into my vein, this non-sterile solution, until my reeling consciousness, nausea, racing heart, and bloated capillaries told me that death was near. Later that night I begged myself to stop. I wrote furiously in my journal, I prayed, I tried to find a logic that would counter the logic of self-destruction. I felt the utter collapse of my life. I left notes for myself: Marc has so much potential. He could be a psychologist, writer, musician, scholar, perhaps all at the same time. But he hasn't got what it takes. I felt the unbearable weight of responsibility for my own life. The weight bore down on my drug-weakened will, a sputtering reflection of splendid futures dying in the listless present.

But the urge would not relent. It grew from its own fulfillment, until, on each occasion, it overpowered every rational thought I thrust against it. My brain was being retuned, of course. It was becoming one-dimensional: dark depression versus bright relief. But I couldn't see it from the inside. I tried allegory. Was it a death wish, a craving for immortality, some monstrous quest for freedom? But the repeated failures at stopping, the letting it go, the one last time, crushed the life out of me until all I could find was self-loathing, utter self-hatred, a wish to stop existing, an end-of-the-world loneliness. Except when the magical passage opened again. And then the loneliness would vanish. As soon as I left the house, got in my car, and drove away to find new treasures, the depression transformed into a sense of power. Because the voice that said *No!* seemed to come from someone else, and my *Fuck you!* freed me from the accusation and scorn that came with it. And I was all right again, long before I found what I was looking for.

How did the impulse materialize and grow so suddenly, with such power? We've seen how craving works in the mind and the brain. We've seen how synaptic narrowing occludes all possibilities except for the one repeatedly practised. We've seen how self-control gets used up by ego fatigue, and how defiance turns that failure into some stilted sense of freedom. But the last ingredient of my addiction—maybe anyone's addiction—was the self-accusation and self-contempt that tormented me, hour after hour, until there was nowhere to go but away. I'm behaving myself: I should feel good. But instead I feel bad and alone. Beside the ragged bodily residue from two nights ago come a host of dark thoughts. And dazed dissatisfaction. The disappointment is crushing. And filling the void, the message of failure blossoms into voices: *The one thing you said you wouldn't do—shoot drugs—you just had to go and do it anyway. Again! And you almost killed yourself. That was two nights ago. And now you're thinking about doing it again?* The scolding voice dredges every ounce of sarcasm, stinking mud from a dead canal. The sarcasm is familiar, well worn. My mother's voice? Not exactly. Whose, then?

If you ask people whether they talk to themselves, or whether they hear voices in their heads, their first reaction will be to try to figure out what you mean. Do you mean literal voices? Do you mean talking out loud? No, just imagined voices. You know they're not really there, yet you hear them. And no, they're not out loud. They're in your head, as people say. Except that sometimes they *are* out loud.

Many individuals report talking to themselves out loud on a daily basis. Yet they're not crazy. Theirs is just a slightly more extreme version of what most or all of us do. And most of us catch ourselves, on occasion, scolding ourselves out loud: "What are you doing? That was dumb!" Or complimenting ourselves out loud: "Way to go!"

When you're the one doing the talking, it's easy to see that it's your voice, not anyone else's. But when the voice is coming at you, when you are the recipient, standing accused, then it's not so obvious whose voice it is. Sometimes it sounds like a parent, or a teacher, or a spouse, or some combination of them. Sometimes it sounds like *you* when you're in a critical mood, scolding someone else, perhaps a child: "Jeez, will you just act your age?" Sometimes it sounds like nobody except that one familiar entity who exists *in your head*—and *only* in your head. And what's the nature of the voice? When you speak words of praise or disapproval out loud, the words are . . . well, they're *words*. But when the accusation (and yes, it's usually an accusation) is coming *at* you, it doesn't always—maybe very rarely—take the form of whole words in whole phrases. It often seems more like a tone of voice than an actual voice. More like the shape or feel of an accusation, a flavour, a *kind* of voice, delivering a *kind* of message. So the whole question of internal voices is hard to unravel.

Until we look into the brain. The OFC and nearby regions of the prefrontal cortex sample incoming information—that's their job—evaluating it quickly and deciphering its emotional significance, then organizing some of the earliest phases of responding to that information: approach, stop, or pull back. In order to do the job properly, the OFC gets its information from the rear of the cortex, where vision and hearing are manufactured, a rich sensory world built of neural firing patterns. That sensory information, sight and sound, gets modified on its way forward. As it travels along the temporal lobes, it gets transformed from pure sensation to meaning. It also gets filtered, combined, and homogenized, so that it becomes a coherent message from the world to you, a meaningful message, telling you something important about the way things are.

The amygdala sits squarely in this informational path, and it connects sensory images to emotional content from the past. Emotional associations in the amygdala are so powerful we can think of them as stamping an imprint on everything that flows through it. All sensory information is coloured with emotion, and that colouring, that meaning, determines what happens next. The amygdala connects forward to the OFC, which elaborates emotional meaning and begins to prepare actions. It connects downward to the hypothalamus, where it fires up ancient response patterns, laid down in the brains of our ancestors at least a couple of hundred million years ago. Patterns like the rage response: baring our teeth and thrusting our bodies forward to repel our enemies. And it connects farther down to the brain stem, where the most primitive emotional urges are orchestrated by tissues that go back at least to reptilian times. It all happens incredibly quickly. It starts within a twentieth of a second. Meaning, memory, associations, instincts, and impulses ignited from the initial flash of perception, waves from the spot where the stone hit the water. All before the cortex has a chance to figure out what's going on.

But here's the thing: the meaning that arises in the temporal lobes, feeding the OFC with news about what to expect and advice about what to do, doesn't necessarily need input. Dreaming is a clear example. Meaning can be nothing more than a limbic memory. It can be practised, elaborated, and crystallized through repetition. So the critical, scolding voices we "hear" are interpretations formed by the OFC— recurring messages of rejection, betrayal, and isolation, based on associations nested in the amygdala and seasoned by a dozen different brain parts. It may not take very much at all—just a rendition of "oops, I did it again" to start the meaning machine, to get it up and running.

And this meaning then recruits all kinds of specifics to bear on our wrongdoing. You *always* do this because you're *weak* or *bad* or *selfish*. That makes sense. That's a coherent story, quickly fleshed out in the prefrontal circuits with the help of more limbic input. But it doesn't come from any specific speaker. It's just an elaborate monologue activated by one or two associations. For me: drugs and pleasure, drugs and danger, drugs and punishment. It's no surprise that these associations recruit much of the brain and organize a seeming *dialogue*, a whole play, in which one role is to criticize and the other is to be criticized.

So here I am, bombarded by this internally generated firestorm of rebuke. My isolation, my memories of recent break-ins, my impulse to try yet another, are plenty of ammunition to get it all going. The rich associations of failure, addiction, and rejection are overripe fruit in my amygdala, colouring the internal voices with scorn. But no actual words are needed. My OFC's job is to anticipate bad things coming this way, and those bad things are accusations that start up before the words are half formed. Yet it keeps coming, and it gets worse. As the night wears on, it gets worse. And that points to one last ingredient: nobody, not even a loathsome addict with zero self-respect, can bear the brunt of this accusation and scorn without getting mad. So add anger to the mix. Anger associations travel downward, to my hypothalamus, where they activate the instinct to fight back, to aggress against this enemy. And then all hell breaks loose. Torrents of neurochemicals, like adrenalin and CRF and a host of peptides, get released into my brain and blood. My sympathetic nervous system gets fired up so that my whole body is taut. Protoplans to attack are instantly launched, to be fleshed out by limbic and cortical agents, waiting upstream. But there's one thing lacking, and that's a victim. Who exactly are we mad at? Who's the perpetrator, the enemy? The loop between my amygdala, OFC, and hypothalamus doesn't wait around for me to figure this out. It's tuned to go right to work. It picks out somebody nearby as the enemy, somebody deserving of violence. It picks out the only person around. And that just happens to be *me*. So now, somewhere deep in my brain, I'm really mad, but the anger just loops back at me: the accusations turn to glaring hatred. The defence turns into more attacks, and the weapons are the almost-heard voices, smears of meaning, left implicit, like poison floating near the bottom of the glass.

The voices don't come from anyone. The voices are patterns of organized, preverbal attack, fuelled by shame and anger that grow each moment, spiralling up from the wounds they themselves inflict. *Are you going to do it again? Go on a greedy, indulgent hunt for pleasure, and then fail to find it through a lousy drug? No wonder you're sitting here on your ass. Oh what shall we do* (in mincing tones)? *What shall we do with ourselves?* And there really is only one thing to do, one way out of this hell: stop being me. Stop being the recipient. Switch places, from passive to active. Let the anger purify itself in the flame of defiance. All I have to say is *Fuck you!* Get in the car, drive away, and I am so far gone, so far away, so not-me, that the critical voices are ashen ghosts scattering in my wake. Now the victim isn't me. The victim is out there in the night.

Somehow I forced myself to stop again. I told myself that I simply could not bear going back to it anymore. It was too strong, I was too weak, and the inevitable depression would gut me until there was nothing left. When temptation tried to sabotage me, I reacted quickly. I turned away, said *Absolutely not!* And sent my mind racing to other chores. Life went back to some semblance of normality.

For a while.

Spring turned to summer, and it was time to begin my internship at Lakehead Psychiatric. I packed my belongings in my VW and drove the twenty hours in two days, stopping at some half-formed town called Wawa to spend the intervening night. The next evening, I drove over a rise and found Thunder Bay spread out below me, a city of a hundred thousand or so, sitting on the back half of Lake Superior. Thunder Bay is frigidly cold in the winter, but it seemed harmless this mild summer night. Yet I felt at once how far away it was from anywhere else I knew. Tina was off in Windsor. She said she loved me, and perhaps she did, but there was no way I could reach out to her from this distance. Still, I had landed a great job, and maybe it was a good thing to be this far from Windsor and Toronto. There would be medical centres and doctors' offices here, too, of course, but I had no idea where they were located, and maybe it would stay that way.

My anxiety was not relieved by the layout of the hospital and the distant corridor to which I was escorted with my bags. The hospital, on the outskirts of the city, consisted of a vast central building meandering over extensive lawns and gardens, with wings or pavilions branching off it every hundred metres or so. All composed of the same red brick. Each wing housed a different population of crazy people: the manic-depressives, the psychotically depressed, the schizophrenics (paranoid and otherwise), and the mentally delayed—or "retarded," as they were called back then. Between the wings were courtyards of a sort—park-like enclosures where the patients could stroll around outside, as long as they didn't leave. My room was in a residence wing for staff, the furthest wing from the main entrance. It was a simple room: a bed by the outer wall with a window above it. A desk, a chair, and a closet. That was it. Just down the hall was a common room where psychiatry residents, psychology interns, and other professional staff might congregate, chat, get to know each other. Except that it was nearly always empty. Where did everyone go at night? All this, though not particularly cheery, seemed survivable. What bothered me the most was that my window looked across the courtyard to a pavilion that housed severely retarded adults, most of whom were classified as psychotic as well. These folks could barely speak, or if they did speak you couldn't understand what they were saying. They could be violent, by accident more than intention, and they experienced tremendous bouts of frustration and despair, expressed through sobbing or wailing sounds, devoid of intention or hope. Most were overweight, and their features, if you got close enough to see them clearly, drooped or sagged in expressions of infantile lassitude, making them look lazy, demanding, or confused, trapped in the deformed bodies of drooling adults. I felt extremely sad for them—sad and a little bit frightened.

The first three weeks went okay. My work was interesting. I was assigned to patients whose mental injuries were immense but whom I thought I might be able to help in some way. At least I could keep them company while they wandered or babbled or told their tales of horror and abuse. Then, at six or so, I'd take my supper in the cafeteria most days. Or I might drive to a restaurant in town. I met a few other interns and they were all nice enough, but nothing really clicked, and I preferred the company of a novel while I ate. I met a staff psychologist who invited me home for dinner with the family. But his generous spirit was an act. There was nobody I could trust behind that smiling face. So I kept mostly to myself, and I was lonely, and the loneliness had its way of invading my resolve. Loneliness came with almost-voices, implicit accusations, that I was not a person anyone wanted to be with. And thoughts of drugs blossomed in these shadows. I was able to resist them for those early weeks. And then something changed. Somehow, all the sadness of the place came crashing in on me, and it fed my own loneliness until I felt suffocated by it. Suffocated and restless. Night after night. And what grew from that turmoil was a sense of hopelessness. Some part of me could no longer believe that I would be able to resist temptation for much longer. And, as any addict knows, that's the ticket to relapse. If it's going to happen anyway, it might as well be tonight.

So it began again. First a street-level window in a private office, not locked tightly enough. Then a medical centre with a dozen offices in an old brick building, each one sealed by a lock that I could violate with my Tupperware tool. Three or four break-ins in a week. That was a lot, even for me. The ground was melting under me and I started to panic. There was no one here I could talk to, no friend, no familiar place, and the howls of the patients in the opposite wing seemed to enter my body and brain while I slept.

The inevitable fall began with a unique odyssey, growing more bizarre by the day. Among the drugs I stole was a large bottle of Desoxyn—oral methamphetamine. People who needed to lose weight were supposed to take one tablet a day. I took three or four when I got back to my room, then mixed it with some opiate and spent the night just below the ceiling, flying through clouds of dreams. When morning came I was still high, but I hadn't slept, and fatigue began to wedge its way into the cracks. So I took about five more. I had a report to write. No problem. A straightforward psychological assessment report. I'd done the tests, calculated the scores, made a bunch of notes. "David's learning difficulties result from cognitive weaknesses in the following areas . . ." Blah-de-blah. This was Psychology 101 stuff. But the whole day seemed to pass in a series of procrastinations. When five o'clock arrived, Dr. Hopkins, my boss, recommended that I take the night off and finish the report tomorrow.

Methamphetamine keeps you high for about eighteen hours. Had I forgotten that? There was just no chance of sleeping that night, so I took more speed, read, played my guitar, and enjoyed myself. The next day, there was so much to talk about. I stopped and chatted with everyone I knew. But when I finally sat down to write, I could tell that my mind was no longer crystal clear. True, I had now missed two nights of sleep. Fatigue was gathering its armies at my gates, but I had the weapons I needed to fortify myself. I'd swallowed more amphetamine that morning, and I felt perfectly balanced, poised to finish the writing. Yet it didn't come easily. I thought maybe I should eat something. I couldn't remember my last meal. But I wasn't hungry, and standing in line at the cafeteria counter it seemed people were looking at me too long, or looking away too quickly. I tried to dim the headlamps of my eyes, which did appear unusually bright in the mirror. But I was sure I was acting okay—in fact, better than okay. I was practically dancing with buoyancy, making irreverent remarks that seemed quite clever. I went to my desk to write again, and I was enchanted with my words. The depths of insight I was achieving were like steel monoliths of truth, painstakingly constructed from obscure pieces of evidence, irrevocably assembled with the cement of logic, rising through the mists of uncertainty as monuments to my razor-like mind. This was going to be a great first report, I could tell.

But I was still not finished at closing time. Dr. Hopkins regarded me steadily. "You can work down here in the Psychology office," he said. "I'll leave a message at the switchboard to lock up when you're done." Then, after a pause: "Just finish it. I'll have Cindy type it up first thing in the morning."

I took more speed and wrote long into the night. It was around 3 a.m. when I was stung by the realization that the purpose of the report was to expand slightly on test results and make recommendations. This should have taken one or two pages, and I was on page 15! Painfully, forcefully, I became aware that I had missed the boat. Not only was I addressing my arguments to some fantasized intellectual forum, but everything I'd concocted was based on the flimsiest of speculation—exactly what I'd been trained not to do in a psych report.

I had to start over, but it was now my third night without sleep.

How was I going to finish by eight-thirty when it had taken all day and half the night to construct this joke of a report? I took more Desoxyn and felt a surge of energy, but less than I'd hoped for. Methamphetamine tolerance builds up fast. I thought I could still pull it off. But when morning came, Hopkins and the secretaries were astonished to find me where they had left me fifteen hours before. I went back to my room with growing concern. Only two paragraphs to go, but I could not think while battling this much fatigue. So I swallowed eighteen more tablets and returned to my task. I ground out one word after another. I tried to make them connect in simple sentences. But each sentence had a mind of its own and floated off into tangential details that had nothing to do with . . . anything. By closing time, Dr. Hopkins looked very concerned.

"What, exactly, is the problem?"

"I just . . . I'm almost done. Really!"

"Marc. I need that report tomorrow morning. The family is coming in for a consultation. Is there something wrong?" The kindness in his eyes had gone dead.

"I was up all night with a fever, Dr. Hopkins. I couldn't sleep. I'd go out like a light if I didn't have my hourly cup of coffee. I'm just not thinking that clearly."

"I just don't see the problem. It's a simple report. A boy with a learning problem."

"Um. I'm sure I can finish it within the hour."

"You just work here until it's done, then. I'm going to ask Cindy to stay behind so she can type it up as you go."

"Thanks, that's—"

"You're still trying to put too much into it, Marc. Just finish it. Make it simple."

I sat at one desk and Cindy sat at another. Other psychology interns came and went, dropping an encouraging word and then moving off quickly, nervously. Each moment seemed to burn with melodrama: images of suspicious disapproval alternating with sympathy—a cheering section. Did anyone still believe the coffee story? My shirt was a sweaty mass of wrinkles. But Cindy waited quietly for each sentence, started typing, and then let me know, as gently as possible, that I didn't seem to be getting to the point. Dream deprivation escalated with whispered suggestions that became almost audible. I knew they were in my head, but I felt that Cindy must hear them, too. Hours went by. The flickering movements at the corner of my eye were getting harder to ignore. Every few minutes my whole visual field seemed to lurch, like a TV channel with poor reception. The dream imagery began to invade every thought, every sentence. And Cindy was now glancing my way with undisguised anxiety.

Then the delusions started in earnest. For at least an hour I kept trying to figure out why Cindy was here, tonight, typing my report, when we'd done the same thing last night. She must be so annoyed with me, having to do it all again. I was utterly convinced that this was the repeat of a previous event. The flat light from the window, the urgency, and above all the humiliation. Exactly like yesterday. Then, with icy shock, I realized this was a déjà vu and I'd gotten lost in it and believed it to be real. I began to see that I could not pull this off. I was collapsing into amphetamine-induced psychosis. Desperation and embarrassment drowned everything else. Cindy tried to help. She would look up, waiting for the next sentence, then read aloud what I'd just written. And I would say, "I've got it," and dictate a sentence whose beginning had misted over by the time it was half-finished. It was hopeless.

It ended when Dr. Hopkins called at 9 p.m. to make sure we were finished. There really were just three or four sentences left to go, but that had been the case for at least an hour. He insisted I leave off for the night, get some sleep, and finish in the morning. I agreed.

I stumbled up to my room, reeling with images of disgrace. They would never look at me the same way again. My reputation was shot. They knew, they must know, that I was on drugs. The cleaning woman last night had known. "I've seen that look before," she'd said. "My son was taking that stuff, but he's all right now." And things had only gotten worse since then. I unlocked my door, faced my empty bed. My head was starting to nod but I would not give in. I was still chasing some symbol of abandon—that speed binge with its built-in lies. The need not to stop, not to sink back, not to accept the commonplace cycle of sleeping and waking that wrecks the illusion of power. I went to my drawer and took out a Benzedrine inhaler (another form of speed)—more booty from my last raid. I broke the container, soaked the material inside it in water for a few minutes, then drank. I had no idea what dosage I was ingesting, or what it would do to me after missing three nights' sleep. But I drank it anyway, and my nervous system jolted like a fallen power line snaking spasmodically on the sidewalk. It was okay. I could handle it. But I needed to take more opiates to balance this crazy energy. I found leftover pills and swallowed them compulsively. I lay down on my bed, waiting for the crawl of sedation . . . finally creeping, creeping up my spine. Here it is, and I'm feeling really fine again. So I go over to the window and I look out. And what I see makes no sense. Across the courtyard, three severely retarded and physically deformed patients stare back at me. They can't be real. But they *are* real. I sneak over to the lounge and look out the window there. They haven't moved; they're still watching me. I'm embarrassed, staring at them like this. But how could they see me? How could they be outside at night? And why are they so big? I go back to my room and peek around the window frame. Monstrous patients, twice the size of normal people, are climbing out of their windows and scaling down the walls, carrying their beds on their shoulders. Down to the courtyard and away to freedom. I watch them with admiration, fear, and a kind of gleeful pride. Who has ever been this high?

I finally did crash that night. Then, for three days, I stayed away from everyone. Dr. Hopkins finished the report. We agreed that I was suffering from nervous exhaustion—a clinical label left over from a hundred years ago. "Just rest," he said. The physical and emotional wreckage left by my binge seemed insurmountable. Depression grew in heaping snowdrifts, obliterating everything. I tried to read—and sleep—for two days. Then the evening of the third day descended slowly and sadly, and I knew without reflection that I would try another break-in. I was strong enough to do it and too depressed not to.

Now it's dark outside. I've been wandering around this enormous clinic for over an hour. I have tried every floor, but the offices yielded almost nothing. And then I find the pharmacy at street level. Its shelves beckon. I can see that there is some kind of motion detector in the ceiling, and yet I walk to the shelves and start to hunt. I tell myself I can get out of here fast. It won't be a problem if I don't linger. I find a large bottle of Tussionex, a few other things, and then I see the lights at the window. Headlights. Still, I'm not panicked. I am moving through molasses, slowly, in a dream. I'm just not getting it. Until I hear doors slam. My chest finally contracts with fear. I get up, bottles in my coat pocket, and run to the opposite end of the building. I hear doors opening and voices shouting. I go down a flight of stairs and find a hall window that looks out into a well against the rear of the building, its top a metre and a half above my head, at ground level. I tell myself I can do this. Up, grab hold of the side, hoist my body through the window, moving now on pure adrenalin. I'm in the well, finally, and I think I might be able to run across the lawn before they find me. I reach my arms up and start to pull myself out. And then I feel large hands on both my arms, helping me up. How kind of you, officers, to help a guy out, a guy in such difficult circumstances. I almost say it out loud. They're not rough, but they're not particularly friendly. They empty my pockets, play their flashlights across my wares, write things on clipboards, handcuff me, and lead me to the waiting car.

At the station, events proceed in assembly-line steps. I'm finger-printed: the thumb and fingers of each hand are clasped, one by one, pressed onto an ink pad, then firmly moved to a series of small squares on a white form. I don't resist. I'm photographed: My head and body are moved into position by another set of hands. I feel very little: a glaze that includes sadness and anxiety, but no real fear. Why aren't I more frightened? Finally I'm escorted to a cell. It's a small room with bars on one wall, just like in the movies. This is where I will spend the night. I sit down on the single cot. This time it's serious. But my thoughts continue to scurry. My feelings shift in little eddies. This time you're an adult, and this is a felony. Then, as the hours drag on, an unexpected peace starts to settle. It had to happen even-tually, and now it has happened. At last it's over.

I called another intern from jail, someone who lived in Thunder Bay, and she called a lawyer. Somehow the lawyer got me released the next day, met me in court, and requested a court date a month away. I found my way back to the hospital, but when I arrived at my room I saw that the door was taped up with yellow police tape. I felt like there was nowhere I could go, and then I realized that everyone here must know. I began to see that getting busted was not going to be my little secret. Humiliation and remorse began to leak from the wound, and that leakage didn't end for months, for years. At times it increased to a torrent. Like when I met with Hopkins later that day. He had little to say, but his eyes were narrowed with disgust. I was told to pack my things and I was escorted back to my room by a security person. Packing took less than half an hour. I looked around at a room completely empty, just as I'd found it, as if I'd never been there at all. And I wished that were true. I closed my eyes and tried to make it true. I wished I had never come.

I stayed at the home of an acquaintance until my court date. I kept my head in my shell; I knew I was not particularly welcome. But there was one thing I had to do, one task to fulfill. My lawyer had been very clear about it during our brief meeting. The only way I was going to stay out of jail was with a heap of glowing character references. I was to call or write everyone I knew, everyone in authority, every supervisor, professor, psychologist, friend in some respected job, and ask each of them to write a letter for me. I'm charged with breaking and entering and theft, you see. I've had this drug problem. And it would be really great if you could say anything nice about me.

And they did. A lot of very nice things, letter after letter. The problem was that now everyone knew. Everyone I'd ever respected and everyone who'd ever respected me. My supervisor at the University of Windsor, Ron Frisch, wrote of the work I'd done, the hard work I'd had to do to get into graduate school in the first place, and what a shame it was that, despite my efforts, my drug problems had remained unresolved. He then phoned to tell me that I was out. Either I left the department of my own accord or he would have me dismissed. I was crushed and I was angry. Although I could see how awkward things might be, I felt I should get another chance. He didn't agree.

"Do you know how priceless our internships are? Do you think we'll ever get another student invited to Lakehead?"

"Yes, but—"

"Besides, you're in no condition to help anyone else deal with their problems. You'd better deal with your own problems before you even *think* of becoming a psychologist." Frisch had a point, no doubt.

My court date finally came around, and the letters must have done the trick. I was sentenced to one year's probation, with an order to attend more therapy. I don't think I would have even minded going to jail for a while. I wasn't sure what else to do with myself. But I went back to my borrowed room, slept for twelve hours, packed, and walked down the block to find some breakfast. The front page of the *Chronicle-Journal* caught my eye from a newspaper box:

> **Brilliant student's future ended by court conviction**
> A Toronto man who showed great promise in the field of psychology now faces a grim future following his conviction on a charge of breaking, entering and theft. . . . Lewis was serving an internship this summer at the Lakehead Psychiatric Hospital (LPH) and life looked very promising. Then on July 21, police officers responded to an alarm . . .

. . . and so on. I couldn't read it all. It made me sick. I forgot about breakfast, found my car, threw in my stuff, and drove away. This was not going to be my little secret at all.

HEALING

You'd think that getting busted, put on probation, kicked out of graduate school, and enduring a kind of infamy that was agonizing to experience and difficult to hide—all of that, and the need to start life over again—would be enough to get me to stop. It wasn't. I returned to Toronto feeling groundless. My life was a wide-open passage leading nowhere. I worked up an alibi to delude my relatives and friends into thinking I'd left school voluntarily. Only a few close friends knew the truth. But of course there were suspicions. Nowhere more pitched than in my department at the University of Windsor. I had no wish to go back there. I had packed up my apartment last spring, and there was nothing more I needed to do.

I joined a co-op house and started looking for work. My housemates were friendly, intelligent people, and I was probably as happy as I could expect to be. A friend and I started a two-man window-washing business. It wasn't as glorious as cruising the halls of academia, but it was a living. We did some painting on the side. I was starting at zero again, but there was some sense of moving forward. Then two things happened at about the same time. I had the audacity to apply for a job in a group home for disturbed children, and that niggling question came up on the application form: Have you ever been convicted of a crime? So I had to tell the story, and of course I didn't get the job. But I got a new painting job. A distant cousin owned a drug store in downtown Toronto, and the ceiling needed painting. I was depressed about my prospects in the mental health field, and so working several days in a pharmacy was probably not a good idea. In the basement, where I mixed the paint and cleaned the brushes, were vast quantities of drugs, mostly old and outdated, but still very potent.

So the cycle started again. And it continued, on and off, for over two years. Periods of resolve, even peacefulness, interspersed by bouts of depression, leading to more break-ins, leading to deeper depression, and so on and so on. I never gave up, though. My efforts to stop continued to diversify, through psychotherapy, journal writing, meditation, talking with friends. And then, in more brutal moments, intense self-flagellation. Once I drank a cup of my own urine, to try to convince myself that I was living off poison. One of the many oaths I wrote to myself was penned in my own blood, to emphasize the extent of my injury. Some friends deserted me, but others remained loyal. They saw that I was struggling with something demonic and they only wished they could do something to help.

My relationship with Tina dissolved into a friendship and I dated women here and there. Then I fell in love—massively, helplessly in love—with a woman named Trish. She was smart, sassy, intensely vulnerable, and very sexy in my eyes. She was also eight years younger than me, and we thought that was funny and sweet. I was supposed to be the mature guy, but I kept no secrets from Trish, at least not at first. She knew my flaws, and they were not unfamiliar to her. Her father was an alcoholic. I was a drug addict. At least I was trying to stop, and often, sometimes for a month or two at a time, I would succeed. But then I would fall off the wagon again, and this hurt her. She didn't get angry or disgusted with me. She got injured. My addiction felt too close to her father's—too threatening, too life-destroying. She worried: about me, especially because I was still stealing drugs, and about us.

Over two years, Trish and I became deeply connected. Her childishness and her capricious moods frustrated me no end, but I loved her with all my heart. I was friends with her friends, she got to know my family, and we talked about moving in together. By now I was living in a small one-bedroom apartment in a stately downtown district. A huge elm tree occupied most of my picture window on the second floor, and a steep staircase led down to a tiny kitchen on the ground floor. I remember Trish in the claw-foot bathtub, soaking her skinny limbs and talking dreamily about having a baby with me some day.

"A Jewish-Irish baby," she imagined out loud. "We'll call him Patrick Moses."

I loved my little apartment, especially because it was full of Trish. She slept there many nights, and many nights I slept at her place. Yet, somehow, Trish was not enough. The cravings would come over me—almost always on nights when she preferred to be alone—and sometimes I could not control them. More drugs, more risks, more damage to both of us. Trish warned me. She didn't know how long she could take it. It wasn't exactly a threat; more like a premonition that this could not last. So I would lie about my drug taking for a few days or weeks, and then I would break down and tell her I had lied. She would listen to me and then move a little farther away. Things became more difficult for us. There were stretches when she didn't want to spend the night together. These only made me long for her more, and that longing scared her and drove her farther away.

When it eventually became clear to me that Trish wanted to "just be friends," my anguish spread like bruising from an injury. I turned away from her, angry and desperate; I went back to drugs with renewed vigour. This culminated in a massive binge: alternately shooting

Demerol and cocaine. I achieved hours of stupor, hallucinations, and delusions. My closest friends found me, semi-conscious, brought me home and put me to bed. I don't know how long I slept, but when I awoke it was to a silent, lifeless world.

Now I find myself lying here alone. I've been awake on and off for hours, and for much of this time I've nursed a very pragmatic fantasy of committing suicide. There are many ways I could do it. I should be a pro by now. Finally, I get up and start pacing. The atmosphere in my apartment is heavy, so thick it takes effort to breathe. Yet laced with some acrid tracery that burns my insides. Images of attack, yet there is no one here. Trish is gone. My friends are gone. The horror on their faces comes alive in memory, and the shame is unbearable. What will I do now? How can this day be survived? I will have to get more drugs. But that will continue the tedious cycle of stupidity yet another week or another two weeks, until I make myself stop for a while. And then I'll start again. And I really can't face it this time. It is just too painful and boring and dead.

Should I call Trish and tell her about my latest binge? Try to elicit some sympathy? It would be like telling her that there was never any chance for us. I could have stayed clean for a month, even two. But for good? What were you thinking, Trish? And it wouldn't matter anyway. Trish is no longer interested.

I shake my head to rid it of the voices, increasingly shrill, accusatory, relentless. And the gloom is so deep that I can barely see through it. I remember the last time she and I spent the day together: there was wind and a Frisbee and leaves swirling magically along the boardwalk. And it all meant nothing. There was no point in any of it. I didn't care about lunch, I didn't care about supper. I didn't care about Trish. Not really. All I cared about was getting more. The old engines had started gearing up below, with their familiar and infinitely boring rattles and groans. A stop at the hardware store. Tupperware. Get out the blade and start carving. Feel the heat of anticipation take control of your limbs. Keep your eyes down. Don't look up. Don't think about it. What would it matter if you did? There's no stopping.

With delicate motions, so as not to tear further, I count my wounds. I take stock. The looks on my friends' faces. That will never go away. It can never be fixed. I can wish and wish and wish and it will never be fixed. And Trish gone. Really gone. It's been over a week, and that's forever. And I have to accept it because I owe her that much. Because I respect her enough to wish her success on her own path, not to subject her to this incessant eroding filth that has overtaken mine. And because I love her that much. For a moment, there's something like relief in it. Then the floodtide of loneliness comes crashing in again. Life without Trish!

Now I sit at my kitchen table and cry. Unfamiliar sounds from inside me. It is so bad. So bad. So bad. And it's all because of drugs. All because I couldn't stop. Because I can't stop. Because I won't stop. And it's killing me. Yes, killing me. And it's not fair.

But something has shifted momentarily. A voice—one of my voices—sounded like it was on my side.

You can't do this to me! I deserve a chance to live. There it is again. This is not the voice I'm used to.

I deserve to be free of you. You can't do this to me! And you won't. I won't let you. I won't . . .

For a moment it's as if someone—perhaps just me—has come to help out.

The words of a past therapist float up from underground: "The strongest thing an existentialist can say is *no.*"

I will say no.

I reflect on this in a slightly different way.

What if I really do . . . ? What if I really really really stop? What if I can't have it back, ever, for the rest of my life? It can't be worse than this. Nothing can be worse than this.

I get up and walk around my kitchen in circles, and I imagine it. I just keep imagining stopping — not for a week or a month, but forever. And I see that it is possible to say *no* every hour of every day. Well, if it's possible, that means I can do it. I repeat this over and over, and the simple logic begins to cohere. Then comes the next step: if I *can* do it, that means I can say I *will* do it. I repeat that over and over as well. Then I shorten it to just four words: *I will do it.* And somewhere in that thought is another breath of warmth, an unfamiliar thaw, a wisp of self-love.

I take a blank sheet of paper from my drawer and I draw a complex shape, somewhere between a flower and a mandala, around the word "No" in the centre of the page. Then I draw lines extending from that nexus to all the edges of the page. So that every bit of the paper is intersected by a line that leads back to the central "No." I work on this for half an hour, embellishing it with elaborate designs that come without effort. My heart is beating slowly, steadily, with a sense of possibility. I dare not think about anything except the goal: to say *no* all day, every day, every moment it's needed. *No* is my friend. *No* can be my centre. I finally look at my work and a slight smile comes to me. That smile is another taste of warmth. That smile soothes me and begins to strengthen me, and I tell myself, Yes, you can do it. You can say no. It's only you who has to be convinced.

I tack my mandala to the wooden frame on the side of the stairway leading up from the kitchen. I take care to centre it at eye level, where I won't be able to avoid seeing it a dozen times a day. I feel strange. This is not quite the same as any other time.

A week later, on a Sunday morning, I toss and turn in bed, currents of opioid depletion like malicious ghosts in my belly and limbs. I don't want to be in the world without opiates . . . by myself, alone. My eyes remain closed and I dig around to find the nearest available dream. I don't want to be awake. Not yet. Not with such a long, empty day ahead of me.

But the ghosts taunt and tease me, with little flares of arousal, stretching me finally into consciousness. And I crawl out of bed, scowling at their persistence. I'm up now. I careen to the bathroom, pee, look at myself in the mirror. Not too bad. Could be worse. And I walk back through my bedroom and into my tiny living room. The tree beyond my window is enormous and majestic. I'm lost in admiration for a moment. Then I return my thoughts to me. What do I do with myself? How do I fill today's void? And then I realize with surprise that the void is not so deep right now. The depression has . . . lifted somewhat: a fog bank receding. Sunlight nearly breaking through. I'm floating through my morning and I'm . . . I'm okay. I'm not fantastic, but I'm okay. There is light here. Not bright, not magnificent, not the sacred light shining from the heart of a pill. But light enough to live, to look around and imagine, without desperation, the day that stretches out in front. Without the constant wounding despair and self-hatred, the suicidal weight of hopelessness. And I think: Has it been that bad? Was it really that bad? And I answer myself: Yes! It really was! *And don't you forget it for one moment.* Because if you do . . . if you dare to minimize, to rationalize, to play, to explore, to knock on that door, even for a lark, even for a short ride . . . you're lost. Again.

Otherwise . . . otherwise, look away. The trance will not claim you if you look away. If you just . . . keep . . . choosing . . . life. You can stay here. It's not so bad, this world, with its subtle shadings, its uncertainties, its tiresome regularities, the sting of loneliness—a mere mosquito bite. Just keep walking forward, and then something else will come.

And then something else. This world . . . means change, uncertainty, boredom, freedom, and possibility. Yes, possibility. That's the one thing that was entirely missing just a week ago. There was only necessity, and something far worse: absolute, utter repetition.

By two or three months it became less difficult. I no longer had to commandeer myself—*Don't even think it!* I no longer had to avoid certain streets, certain districts, where the temptations might start to snowball. I was developing a different sense of myself. I was still a drug addict, no doubt, but now an abstinent one. My one chemical reprieve was alcohol. I allowed myself drinks in the evening—no more than two or three—and it helped provide a structure to my day. A day without a change of state seemed unnatural. Nobody could be that straight. It also gave me something to look forward to when the days seemed long and bleak. Yet it shocked me that, after all these years of addiction, the craving should drop off so rapidly. I no longer had to apply pressure to the wound. It had stopped bleeding.

How could so much neural commitment have broken down so quickly? How could my addled, addicted brain still be that resilient?

I don't actually know the answer. I believe that further research in the neuroscience of addiction will help me get closer to finding it. What I can do is finish my own tale and share a few simple suggestions.

Here's what happened to me.

After quitting for several months, I began a more intensive course of psychotherapy: insight therapy. I stayed with it for many years, and I learned a great deal about shame and sadness, loneliness and anger, and why I sometimes felt compelled to punish and hurt myself. I got a job at a crisis centre for street kids. I was now one of the staff, one of the good guys, and I was helping teenagers who were in almost as much trouble as I'd been, less than a year before. Then, a year later, I reapplied to graduate schools. I was so insecure about my chances of getting back into psychology that I applied to programs that were much less competitive and strenuous—including a Bachelor of Social Work program—as well as the one I really wanted: the child-clinical psychology program at the University of Toronto. For this one, I chiselled my application letter until every word was perfect, and I met with the professor I wanted to work with: Otto Weininger. I told him some of the lurid details of my past life. He didn't seem to be put off. In fact he seemed interested. Two months later I got the official letter: to my amazement, I'd been accepted as a master's student in my first-choice program.

I began with summer courses to get a jump-start on the academic year. I remember the response to my first term paper. My professor, a large, awkward, yet brilliant man, with the unlikely name of Ed Sullivan, came trundling down the hall, waving the paper in the air.

He braked and exclaimed between pants: "I wish there was a grade beyond A+!"

"Are you sure?" I asked. That's all I could think to say.

Over the following months I began to see myself differently: not only as an abstinent drug addict, trying his best, but as a good student with real possibility. That wasn't an easy change after years of self-deprecation. But my term papers continued to elicit compliments. I began to study cognitive development, with a professor named Robbie Case. He was the smartest man I'd ever met, and his theory was a kingdom in which I could reside and flourish, helping to build it, expand it, extend it. Robbie thought my master's thesis was exemplary, and it actually won a prize: the first-place psychology thesis in the province of Ontario. This kind of fame was more palatable than having my name on the front page of the newspaper.

Case became my supervisor for the Ph.D., though I still studied with Weininger every chance I got. I remained fascinated with the emotional dynamics of children and the possibility of helping them. But I loved the world of theory even more, and Case's theory was an evolving edifice of insight and logic that I could carry on, as his apprentice. One day, over lunch, Robbie told me in his typically offhand way that I was born to be a professor. Somehow those words resonated, day and night, for months, until I made up my mind that that was what I wanted. What better life could there be than thinking, writing, researching, and carrying the baton to other students who might follow the trail and advance it further? Just before graduating, I started applying for academic jobs. I went up against some top scholars, and I didn't think my chances were very good. But my specialty—integrating children's cognitive and emotional development— was compelling to the hiring committees. I got a number of offers, and I ended up taking a job at my own department at the University of Toronto.

My past came back to haunt me from time to time. To get licensed as a clinical psychologist, even after two years of internship and practice, I had to explain my conviction record. But it had been a few years now, and I'd done everything possible to put it behind me. I got my licence, and I saw children and adolescents for assessment and psychotherapy. Yet these interests diminished as my pursuit of theory and research took up more of my time. That and family life: I'd gotten married and was now raising a daughter and two stepsons. I wrote long manuscripts for psychology journals, learned to revise them to please the reviewers, and finally started publishing regularly, even though my thinking had moved some distance from that of my mentors. I constructed my own theoretical model and taught it to my students. And I was duly awarded tenure and promoted to associate professor, which meant lifelong job security in a job I liked very much. After nine years of struggling to be a good husband, father, and professor, I was hurt and angered, then simply depressed, by my wife's decision to abandon the marriage. Despair flattened me. But I had my eight-year-old daughter to take care of while finishing a sabbatical in Cambridge. That helped, and I came out in one piece—drug free—a few months later.

After a few more months, I fell in love with my graduate student, Isabel Granic. I was happily obliged to pass the role of supervisor on to another professor, which left Isabel and me free to pursue our non-academic interests. These included travel and martinis, movies, dinners at extravagant restaurants, and playing pool in the local pub, despite the disdainful looks that still came our way occasionally. And intense theoretical debates. Interspersed with lovemaking on her well-worn sofa. Isabel and I were honest with each other in ways neither of us had imagined possible—ways I'd glimpsed with other women, years ago, but that worked so differently now. This time I was whole: there were no distractions.

Finally came promotion to full professor. The year was 2000. With the flexibility this afforded, it was time for a change. I'd been a developmental psychologist for eleven years, but now my attention shifted to the exploding field of emotional neuroscience. Here was a way to study thoughts, feelings, and personality development over the years of childhood, through the workings of the nervous system, the actual flesh that gives rise to all mental processes. It was a route that took me beyond psychological abstractions to the firmer ground of biology, and today I continue to work at the interface of these disciplines.

The spell of drug addiction was broken for me thirty years ago. In the intervening years, I've had my temptations, my flirtations, I've had to be careful with pain medication, but I've never descended anywhere close to the horrific life I lived as a young man. You could say that my life became too full even to consider a return to drugs, but that wouldn't tell the whole story. Not at all. The sculpting of synapses in my early twenties is irrevocable. The meaning of drugs, the imagined value they represent, is still inscribed on my orbito-frontal cortex; and a resonant flair of dopamine can still be ignited in my ventral striatum, at least to some degree. These are the conditions of my nervous system, and they are not reversible. As is well known in the addiction lore, there is no final cure, just recovery, abstention, and self-awareness. But there are happy endings. Mine was a happy ending. It continues to be a happy period of my life. And the brain I'm left with, despite its scars, takes me from day to day as well as—no, better than—I could have expected.

EPILOGUE

IN RECENT YEARS, I've become more and more interested in the neuroscience of addiction, and this book is an expression of that interest. But what have I gained by making connections between the experience of addiction and the brain processes underlying it? This book was never meant to sit in the self-help section, but I continue asking myself: What do I know now that might have helped resolve my difficulties then?

I know that the brain is incredibly sensitive. It has to be, in order to participate in the turbulent flow of reality. But that sensitivity leaves us exceedingly vulnerable to gaps or setbacks in development, especially during adolescence, one of the most chaotic periods of the lifespan. More than that, the impact of events *out there* in the world is paralleled by the impact of events *in here*—in our bodies—and that's because both are manifestations of the same electrochemical processes. So when times are tough emotionally, the stuff that's *supposed to* fulfill children's needs (e.g., getting held by a parent) gets traded in for the chemicals (e.g., opiates) *designed*—both by nature and by the manufacturers—to carry out that fulfillment where it actually takes place, in the flesh sitting smugly behind our fore-heads. Not only that, but the kinds of drugs we seek stand in for the kinds of needs that have gone unfulfilled. Because, from the brain's point of view, both have precisely the same common denominator. The loss of belonging pointed inexorably toward opiates for most of my years as an addict. But when loss of power overshadowed that need, methamphetamine kicked in. As with other brains, mine didn't care where it got the answer, as long as it resolved the immediate question.

I know that the extensive flexibility of our brains is not infinite. On the contrary, it gets used up, relentlessly, permanently, by the sculpting of synapses—a result of experience itself. That's how learning takes place, and addiction is really just a corrupted form of learning. Synaptic flexibility gets used up by the healthy growth from childhood to adulthood *or* by the unhealthy growth of repeating cycles of drug addiction. Synaptic shaping is not only self-promoting and self-reinforcing—an ever-tightening spiral—but it is energized and accelerated by strong emotions. Addictive drugs are addictive *because of* the strong emotions they unleash, emotions that trumpet their meaning and value with increasing emphasis from one occasion to the next. The result is a feedback process: more narrowing of meaning leads to more limits on what feels good, which leads to more constraints on our goals and actions, which leads back to more narrowing. More narrowing in how the world *feels*— from many dimensions down to just one dimension, with a big plus sign at one end and a big minus sign at the other—and a resultant narrowing in how the world behaves toward us. Nobody likes an addict. Not even other addicts. Increasing desperation leads to the only possible source of relief, and that just gives the emotion wheel another spin, carving the ruts ever more deeply.

Feedback is the brain's primary mode of operation. The brain runs on feedback loops in everything it does. But this particular feedback loop is extremely bad news for drug addicts, especially when they cash in at the end of the night, when it comes to the synaptic finale—the overarching pattern of wins and losses that defines who they end up *being*.

And I know that self-control is a tenuous skill, upheld by a population of cells (in the anterior cingulate cortex) that need nourishment, as all cells do. When too much is demanded of those cells, when they have to hold off temptation too long, without the support of an environment, or a relationship, or a philosophy, religion, or narrative sufficient to take up some of the strain, their nourishment runs out and they fail. The life of the addict is a process of dissolution, disintegration: the loss of a sense of self and the loss of a sense of where you fit in the world of other people. Consequently, there isn't much scaffolding to support the wall of the dam; the anterior cingulate has to work against temptation without respite, and the will gives out eventually. The ironic thing is that addicts, known for their lack of strength and resolve, are actually working much harder than anyone should to maintain an extremely difficult balancing act.

I also know that drug addiction is enormously ugly, riding as it does on injury to others and deliberate self-injury, and on trashing the virtues of honesty and self-control we're most likely to cherish. The internal dialogue fed by that ugliness is usually in the red zone, even *before* addiction sets in. Messages of self-rebuke resonate in the orbitofrontal cortex (OFC), which is in charge of anticipating what others think of us and preparing to react. Which means that the enemy is within, and *that* further justifies the logic of self-destruction— another self-reinforcing cycle through which calamities of the mind arise from vulnerabilities of the brain. Yet, because the OFC is more primitive than the engines of language production, and because it comes before them in line, self-disparaging messages don't have to be heard in actual words. Which makes them excessively difficult to pin down and resist. And the same circuits, now bursting with images of shame, elicit fresh raw anger from brain parts that are even more primitive, before they come close to consciousness. So now *rage* rises up into the swell of voices, yet remains under cover, disguised as self-hatred until it switches targets and lashes back at the world instead.

These are some of the insights that occur to me when examining the addicted brain. But if there is a central lesson to be learned, perhaps it's this: The brain's condensation of value is an error. Addiction is a neural mistake, a distortion, an attempted shortcut to get more of what you need by condensing "what you need" into a single, monolithic symbol. The drug (or other substance) *stands for* a cluster of needs: in my case, needs for warmth, safety, freedom, and self-sufficiency. Then it becomes *too* valuable, and you cannot live without it. But one thing cannot be all things. And that's why, in the long run, addictions do such a lousy job of fulfilling needs—if they fulfill them at all. At the same time, many addictions, and certainly addictions to drugs, dash real opportunities to fulfill those needs elsewhere. That's why addictions are such a poor bargain.

Too bad, then, that we cannot remodel our brains, at least not very much and not very easily. Once addiction sets in, the brain will never return to the state—of innocence?—that preceded it. So what is an addict to do? How might I counsel my young self differently, given the neural insights I've collected since then? I suppose there are—still—only a few things that work. Learn to say *no* in a way that can catch and take hold, and support it with a different view of yourself, so that ego fatigue doesn't leach away all your resolve. Fill your life with meanings rich enough at least to compete with, if not defeat, the well-worn synapses of imagined value. Remind yourself that the imagined value is deceptive—that's the way it works. Tune in to what's going on in there. Your brain echoes with messages that can inspire victory or defeat, even when they don't come out in words. Those messages can't be eradicated, but you can add other, gentler voices to the mix. And don't give up. The brain loses a great deal of flexibility with addiction, but it doesn't lose it all.

So my conclusion, based on a neural perspective, is not entirely unique: there are only a few things that can be done to beat addiction, and addicts have to change many parts of themselves in order to carry them out. Yet understanding the brain helps us to understand exactly—or almost exactly—what those things are and why and how they work. That may be the first step of an entirely different approach, not only to the pitfalls of addiction but to the more general pitfalls of being human.

ACKNOWLEDGMENTS

OVER THE FOUR YEARS OF ITS EVOLUTION, this book has had two editors. The first was Isabel Granic, my wife and colleague as I always like to say. She guided, corrected, and inspired me from the earliest outline to the final edits. I'm most grateful to Isabel for her vision — her unwavering conviction that I could use the drama of my own struggles to bring the brain to life for non-scientists. My second editor was Tim Rostron at Doubleday Canada. He helped me understand what a memoir has to *do* in order to invite readers into another person's world, and for that I'm truly thankful.

Michael Levine, my cousin and literary agent, has my warmest gratitude for his unshakable faith in this weird fusion between life and science. I am thankful to other family members as well. My aunts, uncle, and cousins—both Mindens and Lewises—were always encouraging. Their love and confidence bolstered me throughout the writing. My brother and sister-in-law, Michael and Michelle Lewis, read early drafts and stoked my creativity and my nerve with their unstinting support. My father, Allan Lewis, inspired me with his own writing venture. He was the first family member to make a book of his life's story, and that helped encourage me to tell my own.

I want to thank my daughter Zoe for promising not to be too freaked out by tales of her father's drug-crazed youth. And many thanks to friends Tom Hollenstein, Larry Webb, and Alex Lowy, who asked me, year after year, how the book was coming along—and really wanted to know. Rob Firing and Noelle Zitzer guided my steps through the hazards of the publishing world, and they were among the first and strongest advocates for the message of this book. My thanks to the Granics, my in-laws and adopted family, for continuing to believe that this book would someday get written and imagining that it would be worth the wait. And to Ruben and Julian, my two wonderful boys, for all the hours they left me alone to write, and for their bright-eyed wonder that Papa was writing a real book. I'm also grateful to Rebecca Todd and Evan Thompson, fellow academics who saw the value in joining neuroscience with lived experience and encouraged me not to give up on it.

The team at Doubleday Canada lived up to its excellent reputation. Lloyd Davis did a splendid job copyediting the text, and my publicist, Nicola Makoway, launched the book through a remarkable blitz of Canadian media events. The present edition has the support of a marvelous team as well. I want to thank Lisa Kaufman, senior editor at PublicAffairs, for her heartfelt enthusiasm for the book and her relentless efforts to create a jacket design to dress it properly. I'm also grateful to Emily Lavelle, my publicist at Perseus, for pitching it so expertly in the American arena, and to Sandra Beris, the Perseus project editor, for overseeing revisions and additions seamlessly and painlessly.

Finally, I'm grateful to a dear friend who died a few years ago. Through the most intimate conversations, Ron Weinberger helped me recognize the destructive power of addiction in lives other than mine, making it all the more crucial to try to understand it.

ENDNOTES

Chapter 1

p. 7.   Damien Tennant is not his real name.

p. 8.   Tabor Academy still exists, but it is no longer "para-naval" and it is now open to girls as well as boys.

p. 9.   Mr. Witherstein is not his real name.

p. 10   Bob Moore is not his real name.

p. 12. Joe Schwartz is not his real name. Several of the other names used in this section—Carr, Roche, Lavalle, and Miles—have been changed as well.

p. 15. Mr. Wharton's name is not real. The indifference was.

p. 21. Studies of the adolescent brain and its vulnerability to addiction were reviewed by Chambers, Taylor, and Potenza in the *American Journal of Psychiatry*, vol. 160, 2003.

p. 21. For more information on the neuropharmacology of alcohol, see http://www.cerebromente.org.br/n08/doencas/drugs/abuse07_i.htm.

p. 22. Statistics on alcohol use by American teenagers were reported in the National Survey Results on Drug Use, U.S. Department of Health and Human Services, 2001.

Chapter 2

p. 28. The names Raven House and Burdell have been changed. Peter Smits is not the boy's real name.

p. 33. Wikipedia summarizes some of the effects and mechanisms of dextromethorphan intoxication: http://en.wikipedia.org/wiki/Dextromethorphan. Wikipedia also offers a more extended discussion of NMDA antagonists, including dextromethorphan, ketamine, and PCP: http://en.wikipedia.org/wiki/NMDA_receptor_antagonist.

p. 38.   Statistics on the current use of ketamine are from the Substance Abuse and Mental Health Services Administration of the U.S. Department of Health and Human Services.

Chapter 3

p. 43.   All the teachers' names in this chapter have been changed.

p. 52.   The neuronal mechanisms responsible for cannabinoid-induced "plasticity" were discussed in detail by Freund, Katon, and Piomelli in the journal *Physiological Review*, vol. 83, 2003.

Chapter 4

p. 60.   Lisa is not her real name, and White Plains is not where she lived.

p. 65.   The role of dopamine in desire, wanting, or craving has been studied most in relation to addiction. A well-known and comprehensive model of dopamine-based craving was first published by Robinson and Berridge in the journal *Brain Research Reviews*, vol. 18, 1993. For a more recent and more advanced treatment, see Schultz in *Trends in Neuroscience*, vol. 30, 2007.

p. 65.   The role of dopamine (along with oxytocin and other neurochemicals) in love and attachment was reviewed by Sue Carter in *Psychoneuroendocrinology*, vol. 23, 1998.

Chapter 5

p. 82.   The bombardment of the serotonin system by LSD has been a subject of research for half a century. A concise review was published by Aghajanian and Marek in the journal *Neuropsychopharmacology*, vol. 21 (supplement 2), 1999.

Chapter 6

p. 102.   A brain-imaging study of the reactions of white Americans to black faces was published by Cunningham and colleagues in the journal *Psychological Science*, vol. 15, 2004.

p. 114.   Evidence of psychotic states resulting from LSD use was reviewed by Abraham and Aldridge in the journal *Addiction*, vol. 88, 1993.

Chapter 8

p. 132.   The phrase "molecules of emotion" is from a book of that name by Candace Pertt, published by Scribner, 1999.

p. 133.   The enhancement of attachment, bonding, and play by opioids was described by Jaak Panksepp in *Affective Neuroscience*, Oxford University Press, 1998.

p. 134.   Distinct neural systems for "wanting" and "liking" have been described by Berridge and colleagues in many articles, including papers in *Physiology & Behavior*, vol. 97, 2009, and *Current Opinion in Pharmacology*, vol. 9, 2009. Interested readers should visit the (very accessible) Berridge lab website at http://www-personal.umich.edu/~berridge/.

Chapter 9

p. 156.   The neural parallels between starvation and addiction are being studied by John Yeomans, Department of Psychology, University of Toronto.

p. 158.   A key mechanism of addiction is change in glutamate pathways from the orbitofrontal cortex to the ventral striatum (nucleus accumbens), as reviewed by Kalivas and Volkow in the *American Journal of Psychiatry*, vol. 162, 2005. The enhancement of dopamine projections from the VTA to the ventral striatum, as a result of the increase in glutamate release, was reviewed by Kalivas and colleagues in *Neuropharmacology*, vol. 56, 2009.

p. 162.   The concept of an antireward system, and the special role of corticotropin-releasing factor in withdrawal symptoms, were proposed by Koob in the journal *Pharmacopsychiatry*, vol. 42, 2009. This paper can be downloaded from the NIH Public Access archive.

Chapter 10

p. 187.   The dopamine-enhancing effects of amphetamines have been studied for years. See, for example, papers by Ikemoto in *Neuroscience*, vol. 113, 2002, and Clarke and colleagues in *Psychopharmacology*, vol. 96, 1988.

Chapter 11

p. 201.   The James quote is from *The Will to Believe and Other Essays in Popular Philosophy* (1897).

p. 206.   Cortical activation during dreaming has been studied in depth by J. Allan Hobson. See, for example, his recent article in *Nature Reviews Neuroscience*, vol. 10, 2009.

Chapter 12

p. 225.   The basis of drug craving in dopamine flow to the ventral striatum has been reported by many investigators, including Robinson and Berridge in the *Annual Review of Psychology*, vol. 54, 2003. In a particularly clever study, cocaine addicts reported feelings of craving, not pleasure, just before drug administration, when activation of the orbitofrontal cortex and striatum also peaked—Risinger and colleagues, *Neuroimage*, vol. 26, 2005. The connection with glutamate has been described by Kalivas and colleagues in *Neuropharmacology*, vol. 56, 2009.

p. 234.   Sharon is not her real name.

Chapter 13

p. 240.   Dr. Peters and Beth Fenlon are not their real names.

p. 246.   The first studies of ego depletion were by Baumeister and colleagues, published in the *Journal of Personality and Social Psychology*, vol. 74, 1998. Evidence for the neural markers of ego depletion was reported by Inzlicht and Gutsell in the journal *Psychological Science*, vol. 18, 2007, and by Hirsh and Inzlicht in the journal *Psychophysiology*, vol. 47, 2010.

p. 247.   A unified framework for understanding addictive choices, emphasizing cognitive deficits and errors, was published by Redish, Jensen, and Johnson in the journal *Behavioral and Brain Sciences*, vol. 31, 2008.

p. 253.   Bruce Alexander, a Canadian psychologist, published the results of his "Rat Park" experiments in *Pharmacology, Biochemistry & Behavior*, vol. 15, 1981. His findings were so challenging to conventional views of addiction that his research was rejected by higher-profile journals.

Chapter 14

p. 275.   I published a speculative article on the neural mechanisms of internal dialogue in the journal *Theory & Psychology*, vol. 12, 2002.

p. 281.   In *The Night of the Gun* (Simon & Schuster, 2008), David Carr highlights the short circuit between a loss of confidence in the ability to abstain and immediate relapse.

INDEX

Abbie
  letters to/from, 178, 179–180,
    181–182
  relationship with, 170, 171, 175,
    180, 191
ACC. *See* Anterior cingulate cortex;
    see also dACC (dorsal anterior
    cingulate cortex)
Acetylcholine, 80, 154, 204, 205, 206
Acid, 31, 76, 77, 90, 97, 98, 106, 123.
    See also LSD
  amygdala and, 121
  dropping, 96, 99, 110–112, 118
  effects of, 91–92, 111–113, 124
  PCP and, 125
Action, 23, 62, 64, 83, 188
  dopamine and, 64–65
Addiction, 3, 23, 31, 90, 140, 158, 191,
    213, 215, 220, 225, 228, 252, 301
  brain and, 1–2, 154–155, 157, 158–
    159, 226, 302, 305, 306
  drug, 1, 2, 22, 38, 127, 156, 259,
    288, 291, 292, 297, 298, 300, 303,
    304, 305
  fighting, 1, 258, 306
  as a neural mistake, 305
  neuroscience of, 21–22, 297, 302
  physical, 38, 130, 154, 161, 232, 252
  pitfalls of, 306

  psychological, 232, 251
  relationship with, 138–139
  returning to, 272, 292–293, 303
Adolescence, 17, 35, 52–53, 302
Adrenalin, 104, 142, 162–163, 277, 286
Agnew, Spiro, 95
Ah-Kin, 173, 174
Alcohol, 23, 26, 33, 51, 158, 291, 297
  adolescent use of, 22
  brain cells and, 21
  cerebral cortex and, 22
  effect of, 19, 164
  first drink of, 18–21
  GABA transmission and, 24–25
  glutamate transmission
  thinking and, 20
Amphetamine, 51, 187. *See also*
    Methamphetamine
Amygdala, 136, 224, 277
  acid and, 121
  changes in, 154
  emotional associations and, 35–36,
    101, 34, 260–261, 276
  feedback loops and, 226 (fig.)
  freedom for, 121, 124
  high alert of, 107
  norepinephrine and, 119, 163–164
  OFC and, 119–120, 224–225, 227,
    276

Amygdala (*continued*)
  serotonin and, 121
  ventral striatum and, 225
  VTA and, 225, 227
  waking up, 102, 103
  work of, 244–245
Anger, 29, 105, 164, 256, 258, 277,
    288, 292
  anxieties and, 245
  defiance and, 278
  hypothalamus and, 142
  learning about, 297
  shame and, 249, 278
Anhedonia, 68
Anorexia, adolescence and, 17
Anterior cingulate cortex (ACC), 63,
    205, 244, 245 (fig.), 304. *See also*
    dACC (dorsal anterior cingulate
    cortex)
Anticipation, 172, 228, 231, 252
Antihistamines, 183
Anxiety, 27, 89, 101, 112, 133,
    202–203, 212, 224, 250, 265, 279,
    287, 292
  anger and, 245
  depression and, 272
  learning about, 156
  loneliness and, 164
  norepinephrine and, 138
  physiology of, 163
  psychology of, 163
  serotonin and, 121
  shame and, 253
Arrests, 55, 97–98, 121, 218, 287,
    290
Associations, 39, 224
  emotional, 35–36, 101, 260–261,
    276

Balance, Dr., seeing, 272
Bangkok, visiting, 214–215, 216

Barbiturates, 113, 114, 139, 144
Behaviour, 64, 66, 213, 247
  antisocial, 18–19
  ventral striatum and, 224
Behavioural science, 102, 240
Ben
  drugs from, 130
  heroin and, 131, 132, 137, 138, 140,
    141
Benzedrine, 140, 285
Berkeley, 71, 84, 87, 88, 95, 98
  described, 72, 73–75
  living in, 92, 93–94, 106, 128
  meditation in, 197
  returning to, 233
Black guys, problems with, 101–105
Blue cheer, 77
Blue Meanies, 117–118
  See also National Guard; Police
Brain
  addictions and, 1–2, 154–155, 157,
    158–159, 226, 302, 305, 306
  changes in, 26, 154–155, 156
  chemistry of, 21
  complexity of, 3
  drunken, 61
  flexibility of, 303
  image of, 36 (fig.), 62 (fig.), 81 (fig.),
    245 (fig.)
  messages from, 305
  mind and, 228
  physical limitations of, 247
  resiliency of, 297
Brain stem, 33, 36 (fig.), 82, 101, 134,
    136, 206
  fear and, 103
  hypothalamus and, 102
Break-ins, 267, 268, 269, 273, 277,
    286–287, 291
  charges for, 288
  sentence for, 289, 290

Breathing, conscious, 198, 203
Buddhists, 198–199, 214, 215

Calcutta, 148, 215, 229, 248
  leaving, 232, 234
  music scene in, 222
  opium in, 221
  sitar lessons in, 233
  staying in, 220, 221, 222
California Highway Patrol, 97,
  116
Cannabinoid receptors, 52–53, 53–54,
  94
Cannabinoids, 53–54, 78, 80
  firing rate and, 52
  neural etiquette and, 51–52
  neurons and, 51–52
  self-amplification by, 52
  self-made, 51
  synapses and, 73
Carr, Lawrence, 12, 13
Case, Robbie, 298–299
Chemicals, 2, 23, 87, 156, 302
  brain, 51
  inhibitory, 24
Chlorodyne, using, 193
Cocaine, 193, 200, 273
  brain and, 155
  shooting, 293
Codeine, 191, 193, 232
  high from, 180, 258
  stealing, 178
Coincidence detection, 34, 124
Communication, 24, 26, 37
  neural, 23–24
Connie
  Dad and, 148–149
  house of, 165
  Malaysia and, 170, 173, 174–175,
    213
  marriage of, 165

relationship with, 175, 178, 179,
  181, 185, 188
taking money from, 160–161, 164,
  166
Consciousness, 155, 197, 205, 265,
  296, 305
  disturbed, 207
  exploring, 200
  losing, 208
  nature of, 201
  universal, 223
Control, 80
  keeping, 179
  loss of, 247, 260, 262
  See also Self-control
Cortex, 22–23, 33, 34, 62, 65, 101, 134,
  205, 260, 275
  alcohol and, 24–25
  breaking domination of, 39
  change in, 136
  cingulate, 304
  details and, 35
  frontal, 63
  limbic structures and, 36
  lobes of, 36 (fig.)
  motor, 155, 206, 261
  orbitofrontal cortex, 63–65
  prefrontal 62–63
  premotor, 228
  sensory, 39, 206
Cortico-limbic traffic, 37, 38
Corticotropin-releasing factor (CRF),
  162, 277
  adrenalin flow and, 162–163
  rebound, 163
  sympathetic nervous system and,
    182
Cosmic consciousness, 197, 198, 204,
  207
Cough medicine, 2, 27, 232
Crack, brain and, 155

Craving, 1, 61, 97, 228, 252, 266, 272, 292
  brain and, 274
  customized, 226
  dopamine and, 223
  drop off, 297
  meaning and, 225
  mind and, 274
  OFC and, 244
  striatal, 158
  value and, 225, 244
  ventral striatum and, 244
CRF. *See* Corticotropin-releasing factor
Crime, 2, 19, 268, 269, 273, 277, 291
  spiral of, 266–267

D-lysergic acid diethylamide, 80. *See also* LSD; Acid
dACC. *See* Dorsal anterior cingulate cortex; see also Anterior cingulate (ACC)
Dad, 88, 164
  anger from, 166
  Connie and, 148–149
  dating by, 147
  divorce and, 107, 128, 146–147
  drinking with, 176
  drugs from, 180, 189, 214
  family meeting with, 96
  impatience of, 176
  lifestyle of, 149
  love from, 92, 166
  Malaysia and, 170, 173, 174, 213
  marriage of, 165
  Michael and, 148
  relationship with, 174, 178, 179, 181, 182, 183, 184–185, 212
  rescue by, 56–57, 90–91
  smoking with, 147
  taking money from, 160–161

Defiance, 267, 268, 278
Demerol, 183, 263
  shooting, 264–265, 266, 270, 292–293
Depression, 7, 21, 27, 38, 48, 166, 258, 270, 271, 273, 286, 299
  adolescence and, 17
  alternative to, 66
  anxiety and, 272
  combatting, 233, 296
  dopamine and, 66
  heroin and, 136
  impact of, 18
  inevitable, 278
  pooling, 30
  problems with, 59, 291
  psychotic, 279
  serotonin and, 121
  shame and, 30
  smack and, 161
Desire, 35, 61, 157, 252
  addictions and, 156
  dopamine and, 80, 134, 136
  need and, 65
Desoxyn, 281, 283
Despair, 18, 50, 296, 300
Dextromethorphan (DM), 33, 34–35, 38, 43, 51, 52, 53, 123, 204
Developmental psychology, 2, 300
Dexedrine, 265, 266
Dianne (Thomas's girlfriend), 106, 129
Dissociatives, 33, 34, 37, 38, 43, 53, 207, 257, 261
  limbic system and, 205
DM. *See* Dextromethorphan
Doors of Perception, The (Huxley), 95
Dopamine, 2, 64, 67–68, 157, 186–187, 189, 221, 226, 250, 265, 268
  addiction and, 156, 223
  connection and, 188
  feedback loops and, 226 (fig.)

flood of, 65, 97, 154, 158, 160, 187, 223, 224, 248
loss of, 67, 158, 187
lust and, 109
narcotic sedation and, 158
pathways, 226 (fig.)
sensitivity to, 154
Dorsal anterior cingulate cortex (dACC), 244, 245, 245 (fig.), 246–247, 248, 261, 268, 269. *See also* Anterior cingulate cortex (ACC)
limitations of, 247
loss of, 260
Downers, 113
Dreaming, 206, 276,
dream deprivation, 284
Drinking, 25, 28, 33, 60, 61, 161, 272
adolescent, 22
controlled, 297

Easy Rider (movie), 95
Ecstasy, 38
Ego depletion, 246, 261
Ego fatigue, 246, 274
Emotions, 156, 234, 286
amygdala and, 35–36, 101, 34, 102, 260–261, 276
controlling, 246
disorders of, 17
"molecules of," 132
negative, 235
neurons and, 23
Excitation, 23, 24, 81, 185, 248

Fantasies, 27–28, 46–49, 50, 53, 54, 56, 61, 175, 178, 293
Fear, 35, 91, 103, 133, 185, 192, 212, 287
learning from, 156
shame and, 105, 269
Feedback loops, 125, 157–158, 225, 226, 226 (fig.), 303–304

Feelings, 67, 178
meaning and, 36
reality vs., 38
Fight or flight, 101, 162
Firing patterns, 23, 52, 85, 123, 275
Flexibility, loss of, 228, 306
Freaking out, fear of, 91
Fred, 170
Mom and, 169
motorcycle accident and, 171
Freud, Sigmund: cocaine and, 200
Frisch, Ron: dismissal by, 288

GABA, 24, 51, 114, 246
alcohol and, 24–25
Gay men, 99–100, 101, 107
Gelsthorpe, Thomas, 12, 45, 46, 58, 74, 76, 83, 91, 111, 127
advice from, 118, 131
drugs and, 73, 77, 78, 79, 125
girlfriend of, 106, 110, 128
living with, 94, 95
smoking with, 93, 94, 96–97, 117
the Heap and, 123
visiting, 72
Getting
too much, 67
wanting and, 61, 65, 66, 67
Getting down, 160
Ginsberg, Allen, 16, 94
Giving up, joy of, 247–248
Glutamate, 24, 33, 80, 81, 123, 157, 158, 226, 246, 251
alcohol and, 24–25
dopamine and, 159
drugs and, 51
nitrous oxide and, 204
OFC and, 224–225
pathways, 226 (fig.)
slowing/halting, 159

Goals, 35, 135, 188–189, 235, 248, 295
  conflicting, 244
  emotional potency of, 227
  pursuing, 187
Gombak Hospital, 179, 182–183, 184
  described, 176–177
  drugs from, 190–191, 214
Granic, Isabel, 300
Guilt, 67, 145, 191, 256

Hallucinations, 143, 293
Hannah, sex with, 165
Hash, 77
  smoking, 93, 117, 140, 214
Have to, want to and, 65
Heap, the, 118, 123, 127, 135, 152
  described, 123, 129
  drugs and, 130, 131, 137, 138–139
  end of, 144–145
  marriage of, 143–144
  memories of, 243
  problems with, 142
  sexual activities of, 129
  visiting, 126, 128–129, 138, 140
Hedonism, 2, 68
Heroin, 2, 135, 157, 159, 178, 186,
    216, 218, 242
  brain and, 155
  buying, 151, 217
  depression and, 136
  dreaming about, 179
  effects of, 132, 142–143, 152–153
  high on, 153–154, 172, 217
  overdosing on, 140–142, 144
  receptor sites and, 136
  red rock and, 219
  scoring, 130, 150, 154, 219
  using, 131–132, 136–139, 143, 149,
    152–153, 185, 232
  withdrawal symptoms from, 161–162

Highs, 51, 78, 97, 112, 142, 150, 155,
    184, 199, 286
  codeine, 180, 258
  heroin, 153–154, 172, 217
  methamphetamine, 268, 281–282
  searching for, 220
Hippies, 89, 95, 115, 119, 120–121
Hippocampus, 35, 124, 154, 155, 206,
    245, 261
Hitchhiking, 97, 100, 101–105, 110
Homosexuals, 99–100, 101, 107
Hopkins, Dr., 281, 282, 284, 286
  concern from, 283
  meeting with, 287–288
Howard Johnson's restaurant, arrest at,
    53–54, 55
Huxley, Aldous, 95
Hydrocodone, 257
Hyperarousal, 161, 186, 269
Hyperrealism, 186, 209
Hypothalamus, 36 (fig.), 101, 133, 162,
    206, 276, 277
  anger and, 142
  brain stem and, 102
  changes in, 154
  fear and, 103
  norepinephrine and, 164
  opioids and, 132
  peptides and, 233

Identity, 99, 106, 128
  damage to, 16–17
  searching for, 74
Impulses, 276
  controlling, 246, 261
Independence, 15, 72, 179
Indian Room, 147, 149, 166, 221
Information, 54, 80, 83
  LSD and excess, 81, 83, 113, 118
  incoming, 107

sensory, 275
serotonin and, 81, 83
Inhibition, 64
neuronal, 23–25
Inner world, exploring, 193
Insight therapy, 297–298
Internal dialogues, 128, 182. *See also*
Voices, internal
Isolation, 240, 253, 277

Jail, 90
acid trip in, 91
release from, 287
time in, 55–56, 98, 121, 289
James, William, 200, 201, 205, 206, 207
Jim
death of, 144
described, 122–123
heroin and, 139, 141
marriage of, 143–144
PCP and, 125
Pumpkin and, 129
Ralph and, 129, 139, 144–145, 243
Jimmy, 154, 160, 220
departure of, 161, 164
falseness of, 152
hanging out with, 159, 172
heroin and, 153, 232
meeting, 149–150
Joints, swallowing, 98
Junkies, 130, 131, 150, 151–152, 248,
249

Ketamine, 33, 34, 38, 123, 124
Krishnamurti, 176, 180, 181, 199
Kuala Lampur, 175, 179–180, 193,
194, 216
arrival in, 172–173
leaving, 212
Kubrick, Stanley, 71

Lakehead Psychiatric Hospital
dismissal from, 288, 289
internship at, 272, 278–279
Laos, visiting, 215, 216, 217–219
Laura, 165, 174, 185
Learning, 283, 303
corrupted, 303
synapses and, 155, 156
Lebanese Blonde, smoking, 94
Leonard (marriage counselor), seeing,
256–257, 260, 262
Libido, 108, 160
Liking, wanting and, 134, 135
Limbic system, 35, 36–37, 36 (fig.), 40,
62, 124, 136, 204, 261, 277
cortex and, 36
dissociatives (DM, nitrous oxide)
and, 36–37, 39, 205
meaning and, 35–38
Lisa, 134
being with, 60, 68, 97
dopamine and, 66
drinking with, 60, 61
wanting/getting, 65, 66–67
Loneliness, 106, 133, 154, 191, 240,
273–274, 280, 294
addiction and, 233
anxiety and, 164
learning about, 297
persistent, 127
sting of, 296
Love, 146, 166, 180, 256
addiction to, 1
looking for, 108
LSD, 2, 80, 97, 125, 137. *See also* Acid;
D-lysergic acid diethylamide
effects of, 84–86, 87–88, 90,
91–92
explaining, 91–92
overdose of, 113

LSD (*continued*)
serotonin and, 82, 84, 107, 119
taking, 58, 78, 82, 84
Lysergic acid, 90. *See also* LSD

Malays, 172, 176, 177, 194, 195,
210
Malaysia, 193, 194, 214, 216, 221
leaving for, 170, 171, 172
Manic depression, 270, 279
Marijuana, 77, 104, 221
preoccupation with, 54
shunning, 57, 60
smoking, 48–49, 50, 52, 53–54, 71,
73, 78, 94, 96, 98, 147
Meaning, 35, 227, 228, 275, 276, 278
craving and, 225
disconnected, 204
feeling and, 36
limbic basis of, 37, 38, 40
reality and, 38–39
sense and, 35–38
Meditation, 197, 198–199, 291
Memory, 39, 155, 260, 276
frightening, 164
purging, 105
Mescaline, 96, 138, 140
Meth-opiate cocktail, 268, 285
Methadone, 180, 184, 189
Methamphetamine, 184, 185, 190,
265, 303
effects of, 155, 186–187, 187–189
high from, 268, 281–282
taking, 186, 282
tolerance for, 283
Methedrine, 184
Michael, 71, 96, 106, 147, 153
comfort from, 234
Dad and, 148
help from, 113
lifestyle of, 149

playing/singing with, 146
tabla and, 233
Mind
brain and, 228
craving and, 274
Modern Lodge, staying at, 220, 221
Mom
boyfriend of, 147
discontent and, 268
divorce and, 107, 128, 146–147
family meeting with, 96
Fred and, 169
gaze of, 213, 214
motorcycle ride with, 170–171
relationship with, 170, 171, 178–
179, 180
Mood change, brain change and, 21
Morphine, 135, 183, 242, 245, 248,
257
addiction to, 251–252
thinking about, 243
Motivation, 228, 260, 262
Motorcycle, 146
accident with, 170–171
arguments over, 234
riding, 169–171
Music
comfort from, 234
Indian classical, 221

Narcotics, 180–181, 184, 242
National Guard, fear of, 117
Needs
addictions and, 305
desire and, 65
Nervous system, 3, 22, 247, 285
Neurochemicals, 101, 134, 135, 156,
204, 206, 228–229
drugs and, 1
flow of, 3
torrents of, 277

Neuromodulators, 2, 64, 81, 163, 185
  changes in, 154
  dopamine and, 80
  sketch of, 81 (fig.)
Neurons, 25, 34, 37, 89, 94, 136, 206, 246
  cannabinoids and, 51–52
  dopamine and, 157, 244
  emotions and, 23
  flooding, 83
  function of, 22–23
  information from, 54
  opioids and, 134
  serotonin and, 81, 82
  synchronizing, 124
  threshold of, 80
Neuropeptides, 132
Neuroscience, 2, 3, 228, 300
Neurotransmitters, 23–24, 52, 64, 246
  drugs and, 51
  excitatory, 23–24, 80–81
  inhibitory, 23–24, 114
Nietzsche, Friedrich, 45
Nitrous oxide, 2, 206, 207, 214, 265
  effects of, 202–204
  glutamate and, 204
  limbic system and, 205
  using, 199–200, 200–201, 202
NMDA antagonists, 123, 124, 204
NMDA receptors, 33, 37, 52, 123, 124, 204, 205
  dextromethorphan and, 34–35
  shutting down, 39, 125
No, saying, 295
No-no, described, 129–130
Norepinephrine, 80, 101, 107, 121, 187, 265, 268
  amygdala and, 163–164
  anxiety and, 138
  hypothalamus and, 164
  serotonin and, 119

Obsessive-compulsives, dopamine and, 64
Occipital lobes, 37, 82
OFC. See Orbitofrontal cortex
Opiates, 51, 138, 160, 193, 233, 257, 281, 302, 303
  external, 253
  living without, 296
  semi-synthetic, 172
  sedation from, 285
  source of, 162
  using, 191, 219, 221, 267
Opioid receptors, 134, 136–137
Opioids, 136, 162
  depletion, 296
  dopamine and, 135
  evolution of, 133–134
  hypothalamus and, 132
  internal, 227, 253–254
  security and, 232–233
  using, 193, 220, 254
Opium, 2, 134, 223, 225, 232
  addiction to, 232
  raw, 221
  refined, 221, 231
  using, 140, 221, 233–234
Orang Asli, 176, 177, 178, 194, 210, 211
Orange wedge, 76
Orbitofrontal cortex (OFC), 65, 67, 89, 102, 159, 172, 248, 251, 275, 301, 304
  amygdala and, 119–120, 225, 227, 276
  changes in, 157, 158, 261
  craving and, 244
  dopamine and, 225
  feedback loops and, 226 (fig.)
  glutamate and, 224–225
  interpretation by, 103
  ventral striatum and, 64, 158

Orbitofrontal cortex (*continued*)
  ventral tegmental area (VTA) and,
    225, 227
  work of, 63–64, 244–245, 277
Orbitofrontal neurons, 157, 159
Orbitostriatal circuits, 64, 68, 158, 227
Outer world, exploring, 193
Overdosing, 113, 140–142, 144
Oxycodone, 172

Pain, 256, 260
  caring for, 133, 172
  feeling, 134, 290
  medications for, 172, 180, 181, 300
Panic, 30, 48, 86, 97, 98, 113, 120, 121,
    191, 239, 271, 286
Paraphernalia, 18, 75, 155, 159, 185
  affection for, 94
  possession of, 98
Parkinson's disease, dopamine and, 64
Paxil, 81
PCP. *See* Phencyclidine
Peace Corps, 176, 194
Penny Saver supermarket, shopping
    trips to, 93, 94, 95
People's Park, 110, 115, 116
Peptides, 2, 132, 233, 277
Perception, 23, 125, 188, 204
Percodan, 172, 178, 180, 257, 262, 265
Perry, 12, 16, 34, 38, 45
Peters, Dr., 235, 240
Phencyclidine (PCP), 38, 123, 204
  throbbing of, 126
  taking, 124–125
Plasticity, 33, 52, 227
Pleasure, 63, 137, 277
Police, 92, 94, 107
  arrest by, 55, 98, 287demonstrations
    and, 117
  encounter with, 90, 120–121
  fear of, 118

forgetting about, 126
hippies and, 89, 119
Peoples' Park and, 116
Pond House, 12, 16, 45
Prefrontal cortex, 62, 64, 82, 87, 102,
    134, 244, 277
  changes in, 154–155, 157
  dorsal-lateral, 85, 206, 245, 261
  work of, 63
Prescriptions, forging of, 257
Probation, 56, 289, 290
Probation officer, visiting with, 57,
    59–60, 66
Prostitute, night with, 218
Prozac, 81
Psilocin, 140
Psilocybin, 96, 138, 140
Psychedelics, 97, 123, 127, 140
  destructive power of, 110–111
  fascination with, 96
Psychotherapy, 291, 297–298, 299
Psychotics, 114, 280
Psychotropics, 75
Puerto Rico fantasy, 46–49, 50, 53,
    54, 55
Pumpkin, 126
  described, 122
  Jim and, 129
  marriage of, 143–144
  overdosing and, 141
  PCP and, 125
  Ralph and, 129, 130, 139
Purple Haze, 77
Purples, 77

Racist asshole, accusations as, 101, 102,
104, 105
Ralph, 123, 126, 128–129
  described, 122
  heroin and, 130, 131, 136, 137, 138,
    139, 140, 141, 142

Jim and, 129, 139, 144–145, 243
marriage of, 143–144
mental institution for, 144
overdosing and, 141
PCP and, 125
Pumpkin and, 129, 130, 139
Rapid eye movement (REM), 205–206
Rats, 235, 245, 247, 248
experimenting on, 240–241, 242, 243, 250–251, 253
Reagan, Ronald, 94, 115
Reality, 35, 54, 79, 84, 207, 242
access to, 201
coming back to, 114
meaning and, 38–39
turbulent flow of, 302
Receptors, 2, 23, 82, 134
heroin and, 136
opioids and, 136
Recovery, 2, 172, 179, 258, 301
Red rock, smoking, 219
Reds, 77
teasing by, 10–11, 12
Remorse, 161, 262, 287
Restlessness, 106, 128, 221
Reward, 63, 67–68, 73
Rice, Anne, 242–243
Romilar, abusing, 30, 31–33
Roy, Hiren, 222

Sadness, 287
learning about, 297
Safety, 166, 267, 305
San Francisco, 96, 165
living in, 71–72, 128, 147–148, 160
Schizophrenia, 34, 124, 279
Schwartz, Joe, 12, 34, 45, 46, 57, 66
arrest and, 55, 56
switching schools and, 47, 48, 49, 53
Seconal, 113, 139, 243. See also reds
Security, 180, 181, 232–233

Sedation, 113, 153, 164, 190
opiate, 162
Self-awareness, 176, 301
Self-consciousness, 85, 87, 211
Self-contempt, 27, 68, 164, 190, 274
Self-control, 90, 201, 206, 244, 250, 252, 274, 304
loss of, 85, 247, 248, 260
Self-criticism, 15, 25, 57, 105, 273, 274, 291, 298
Self-destruction, 235, 269, 273, 304
Self-doubt, 190, 233
Self-hatred, 273, 296, 305
Self-help, 176, 302
Self-image, 211, 253
Self-improvement, 128–129
Self-reinforcement, 303, 304
Self-regulation, loss of with LSD, 82–83
Selfhood, 17, 61
Sense of self, loss of, 260, 304
Serotonin, 90, 113, 124, 187, 268
amygdala and, 121
depletion of, 125
depression and, 121
excitation and, 81
information flow and, 80, 83
LSD and, 82
neurons and, 81, 82
norepinephrine and, 119
synapses and, 97
Serotonin receptors, 80
LSD and, 84, 107, 119
Sex, 110, 127, 130, 158, 165, 231
addiction to, 1
dopamine and, 64
friendship and, 146
looking for, 107, 108–109
smack and, 160
Shame, 57, 68, 100, 142, 178, 191, 250, 285

Shame (*continued*)
  addiction and, 233
  anger and, 249, 278
  anxiety and, 253
  depression and, 30
  fear and, 105, 269
  guilt and, 67
  images of, 304
  learning about, 297
Shankar, Ravi, 147
Sharon, 241, 242, 247, 266
  addictions and, 252–253, 255, 258
  arguments with, 234–235, 245,
    249–250, 253, 256, 258, 259–260,
    267, 268, 270–271
  attachment to, 254, 269
  leaving, 255–256, 259–260, 262–
    263, 271–272
  problems with, 254, 255
Sheila, relations with, 110
Sitar
  buying, 221–222, 223, 232
  studying, 233
Smack, 144, 149. *See* Heroin
  depression and, 161
  libido and, 160
  scoring, 150, 154, 217
Speed, 78, 184. *See also* Amphetamine,
    Methamphetamine
  using, 185–196, 285
Sproul Plaza, 75, 86, 93
  demonstrations at, 77–78, 116,
    117–118
Stealing, 2, 190–191, 214, 262, 263–
  264, 268, 273, 292
Stillwell, Mr., 43
STP, 77, 96
Stress, 134, 163
Striatum, 62 (fig.), 136
  anticipation and, 172
  changes in, 154

OFC and, 64
  ventral, 63, 64, 97, 158. *See also*
    Ventral striatum
  VTA and, 187
Subcortical structures, 63, 206
Suicide, 17, 273, 293
Sukert, Mr., confrontation with, 41–42
Sullivan, Ed: course with, 298
Susan, 128, 146
  dropping acid with, 110–112
  finding, 107–110
  Peter and, 108–109, 110
  psychedelic drugs and, 110–111
  sex with, 108–109, 127
Switching schools, fantasy about,
  46–49, 50, 53
Sympathetic nervous system, 101, 103,
  104, 162, 182
Synapses, 2, 23, 34, 65, 81, 155, 157,
  187, 223, 226, 267
  active, 227
  cannabinoids and, 73
  dopamine and, 67, 224
  flexibility of, 303
  learning and, 155, 156
  sculpting, 227, 228, 300–301, 303
  serotonin and, 97
  strengthening, 227
  striatal, 220–221
  well-worn, 274, 305

Tabor Academy, 43, 56, 61, 93, 96
  leaving, 27, 47, 48, 66, 72, 73, 76,
    213
  life at, 7, 8, 9–10, 12, 16, 17, 26, 44,
    50, 57, 58
  problems at, 17–18, 59, 105
  dealing with, 30–31, 58
  fight with, 28–29, 58
  first drink with, 18–21
Teasing, 10–11, 12–13, 14

Telegraph Avenue, 84, 85, 88, 115, 119
    described, 74–75, 76
    shopping on, 93
Temporal lobes, 37, 82, 275, 276
Terry, 196–197, 201, 209–210
    consciousness and, 200, 208
    meditation and, 197–198, 199
    nitrous oxide and, 199, 203
Thailand, visiting, 214–215, 219–220
Thalamus, 36 (fig.)
Tina, 143, 272, 279, 291
Toronto, 255, 259
    returning to, 256, 273, 290
    summer in, 233–234
    work in, 291
Tranquilizers, 183, 184, 263
Trish, 291–292, 293, 294
Truck driving, 148, 149
Tussionex, 257–258, 273, 286

University of California, 75, 95, 115
    dropping out of, 146
    Peoples' Park and, 116
University of Toronto, 299
    child-clinical psychology program
        at, 298
    undergraduate work at, 234
University of Windsor, 255, 288, 290
USAID workers, 194

Value, 225, 305
    craving and, 225, 244
    orbitofrontal cortex (OFC) and,
        158, 224–225
Ventral striatum, 134, 135, 159, 228,
    248, 251, 301

amygdala and, 225
behaviour and, 224
craving and, 244
dopamine and, 187
feedback loops and, 226 (fig.)
OFC and, 158
work of, 244–245
Ventral tegmental area (VTA), 135,
    157, 224, 251
    amygdala and, 225, 227
    feedback loops and, 226 (fig.)
    OFC and, 225, 227
    (ventral) striatum and, 187
Vientiane, 315
    visiting, 216, 217–219
Vietnam War, 75, 77–78, 110, 216
Violence, 106, 127, 278
Voices, internal, 136, 191, 274–278,
    280, 293, 294–295
VTA. See Ventral tegmental area

Wanting, 61, 97, 156
    getting and, 61, 65, 66, 67
    liking and, 134, 135
Warhol, Andy, 94
Weininger, Otto, 298–299
Windsor, 258, 262, 272, 279
Withdrawal symptoms, 64, 161–162,
    181, 223, 232, 254, 256, 269,
    271
Witherstein, Mr., 9, 43
Working memory, 63, 204, 206, 244,
    245, 261

Zoloft, 81

ABOUT THE AUTHOR

Duncan de Fey

Dr. Marc Lewis is a developmental neuroscientist and professor of developmental psychology, recently at the University of Toronto, where he taught and conducted research from 1989 to 2010, and presently at Radboud University in the Netherlands. He is the author or co-author of over fifty journal publications in neuroscience and developmental psychology.

PublicAffairs is a publishing house founded in 1997. It is a tribute to the standards, values, and flair of three persons who have served as mentors to countless reporters, writers, editors, and book people of all kinds, including me.

I. F. STONE, proprietor of *I. F. Stone's Weekly*, combined a commitment to the First Amendment with entrepreneurial zeal and reporting skill and became one of the great independent journalists in American history. At the age of eighty, Izzy published *The Trial of Socrates*, which was a national bestseller. He wrote the book after he taught himself ancient Greek.

BENJAMIN C. BRADLEE was for nearly thirty years the charismatic editorial leader of *The Washington Post*. It was Ben who gave the *Post* the range and courage to pursue such historic issues as Watergate. He supported his reporters with a tenacity that made them fearless and it is no accident that so many became authors of influential, best-selling books.

ROBERT L. BERNSTEIN, the chief executive of Random House for more than a quarter century, guided one of the nation's premier publishing houses. Bob was personally responsible for many books of political dissent and argument that challenged tyranny around the globe. He is also the founder and longtime chair of Human Rights Watch, one of the most respected human rights organizations in the world.

• • •

For fifty years, the banner of Public Affairs Press was carried by its owner Morris B. Schnapper, who published Gandhi, Nasser, Toynbee, Truman, and about 1,500 other authors. In 1983, Schnapper was described by *The Washington Post* as "a redoubtable gadfly." His legacy will endure in the books to come.

Peter Osnos, *Founder and Editor-at-Large*